AQA GCSE

MATHEMATICS

Foundation sets

PRACTICE BOOK

Series editor: **Glyn Payne**

Authors: **Greg Byrd, Lynn Byrd**

www.pearsonschools.co.uk

✓ Free online support
✓ Useful weblinks
✓ 24 hour online ordering

0845 630 22 22

Part of Pearson

Longman is an imprint of Pearson Education Limited, a company incorporated in England and Wales, having its registered office at Edinburgh Gate, Harlow, Essex, CM20 2JE. Registered company number: 872828

www.pearsonschoolsandfecolleges.co.uk

Longman is a registered trademark of Pearson Education Limited

Text © Pearson Education Limited 2010

First published 2010
14 13 12
10 9 8 7 6 5 4 3 2

British Library Cataloguing in Publication Data
A catalogue record for this book is available from the British Library.
ISBN 978 1 408 23273 6

Copyright notice
All rights reserved. No part of this publication may be reproduced in any form or by any means (including photocopying or storing it in any medium by electronic means and whether or not transiently or incidentally to some other use of this publication) without the written permission of the copyright owner, except in accordance with the provisions of the Copyright, Designs and Patents Act 1988 or under the terms of a licence issued by the Copyright Licensing Agency, Saffron House, 6–10 Kirby Street, London EC1N 8TS (www.cla.co.uk). Applications for the copyright owner's written permission should be addressed to the publisher.

Edited by Lauren Bourque
Designed by Pearson Education Limited
Typeset by Tech-Set Ltd, Gateshead
Original illustrations © Pearson Education Ltd 2010
Illustrated by Tech-Set Ltd
Cover design by Wooden Ark
Cover photo © Corbis/NASA/JPL-Caltech
Printed in Malaysia, CTP-KHL

Acknowledgements
Every effort has been made to contact copyright holders of material reproduced in this book. Any omissions will be rectified in subsequent printings if notice is given to the publishers.

BLACKBURN COLLEGE
LIBRARY
Acc. No. BB59061
Class No. 510.76 PAY
Date JUN 14

Quick contents guide

2010 Specification changes: Assessment Objectives and functional maths

From 2010 the AQA GCSE Maths specifications have changed. For both Modular and Linear, the main features of this change are twofold.

Firstly the Assessment Objectives (AOs) have been revised so there is more focus on problem-solving. The new AO2 and AO3 questions will form about half of the questions in the exam. We provide lots of practice in this book, with AO2 and AO3 questions clearly labelled.

Secondly about 30–40% of the questions in the Foundation exam will test functional maths. This means that they use maths in a real-life situation. Again we provide lots of clearly labelled practice for functional questions.

What does an AO2 question look like?

"**AO2** select and apply mathematical methods in a range of contexts."

An AO2 question will ask you to use a mathematical technique in an unfamiliar way.

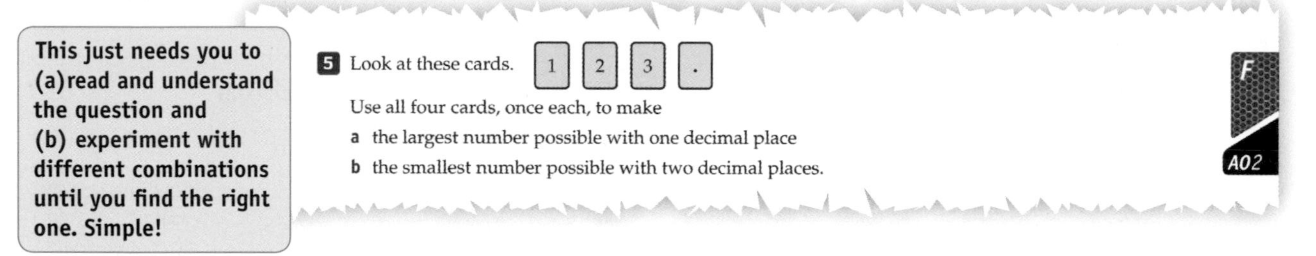

This just needs you to (a) read and understand the question and (b) experiment with different combinations until you find the right one. Simple!

5 Look at these cards. 1 2 3 .

Use all four cards, once each, to make

a the largest number possible with one decimal place

b the smallest number possible with two decimal places.

F AO2

What does an AO3 question look like?

"AO3 interpret and analyse problems and generate strategies to solve them."

AO3 questions give you less help. You might have to use a range of mathematical techniques, or solve a multi-step problem without any guidance.

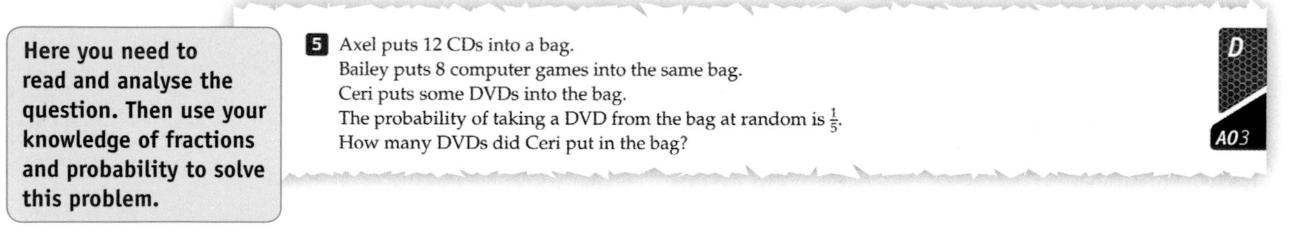

Here you need to read and analyse the question. Then use your knowledge of fractions and probability to solve this problem.

5 Axel puts 12 CDs into a bag.
Bailey puts 8 computer games into the same bag.
Ceri puts some DVDs into the bag.
The probability of taking a DVD from the bag at random is $\frac{1}{5}$.
How many DVDs did Ceri put in the bag?

D AO3

What does a functional question look like?

When you are answering functional questions you should plan your work. Always make sure that you explain how your answer relates to the question.

G

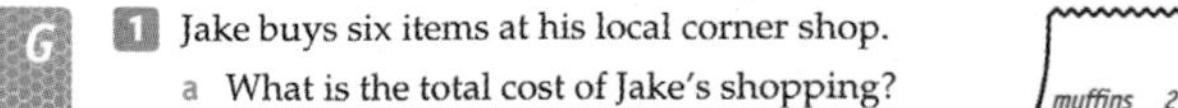

1 Jake buys six items at his local corner shop.

a What is the total cost of Jake's shopping?

b Jake pays with a £10 note.
How much change will he receive?

Read the question carefully.

Think what maths you need and plan the order in which you'll work.

Follow your plan. Check your calculation. Job done!

How to use this book

This book has all the features you need to achieve the best possible grade in your AQA GCSE Foundation exam, **both Modular and Linear**. Throughout the book you'll find full coverage of Grades G-C, the new Assessment Objectives and Functional Maths.

At the end of the book you will find a **complete set of Practice Papers for Modular and a complete set for Linear**.

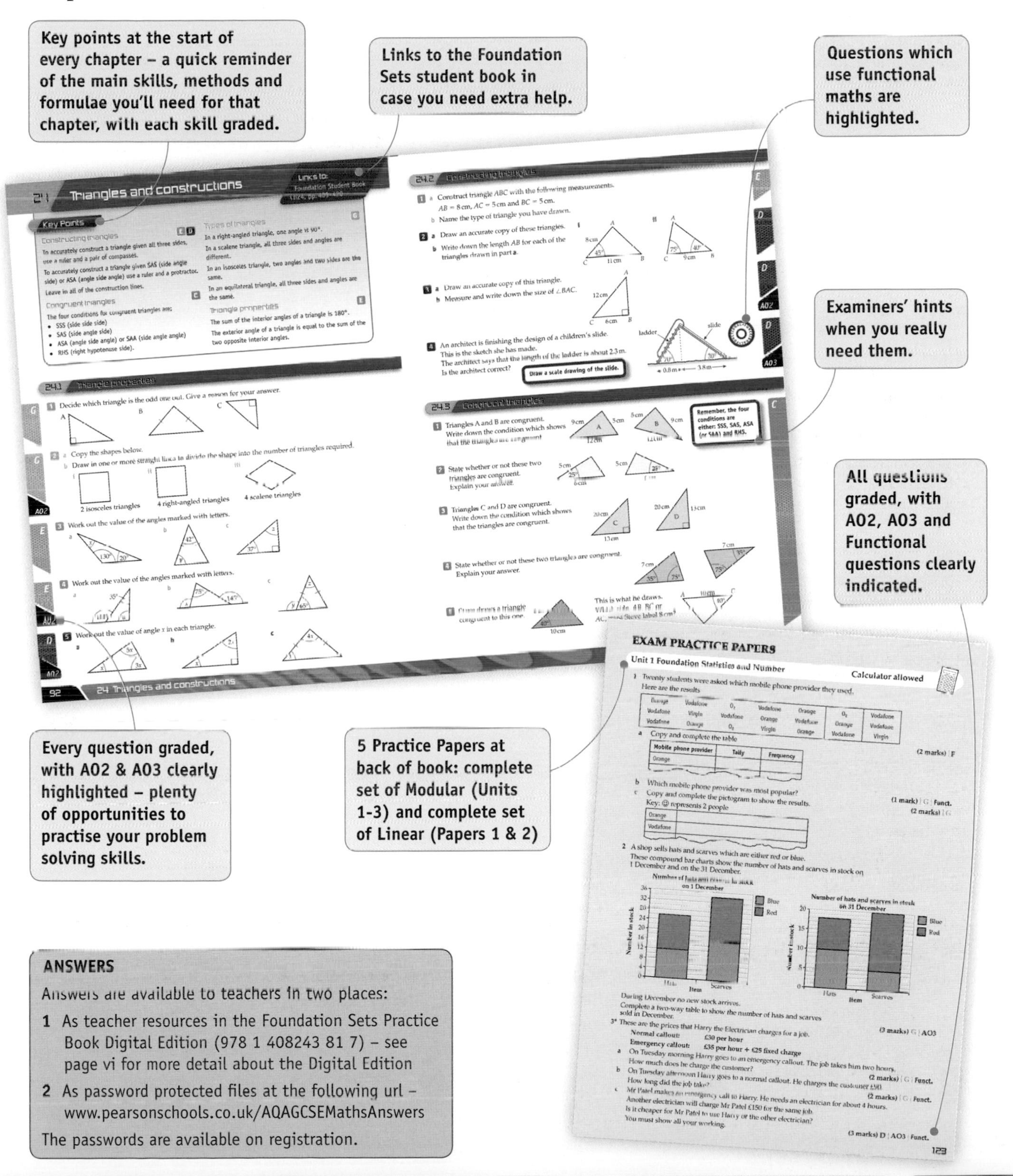

ANSWERS

Answers are available to teachers in two places:

1 As teacher resources in the Foundation Sets Practice Book Digital Edition (978 1 408243 81 7) – see page vi for more detail about the Digital Edition

2 As password protected files at the following url – www.pearsonschools.co.uk/AQAGCSEMathsAnswers

The passwords are available on registration.

About the Digital Edition

We have produced a **Digital Edition** of the Foundation Practice Book (ISBN 978 1 408243 81 7) for display on an electronic whiteboard or via a VLE. The digital edition is available for purchase separately. It makes use of our unique **ActiveTeach** platform and will integrate with any other ActiveTeach products that you have purchased from the **AQA GCSE Mathematics 2010 series**.

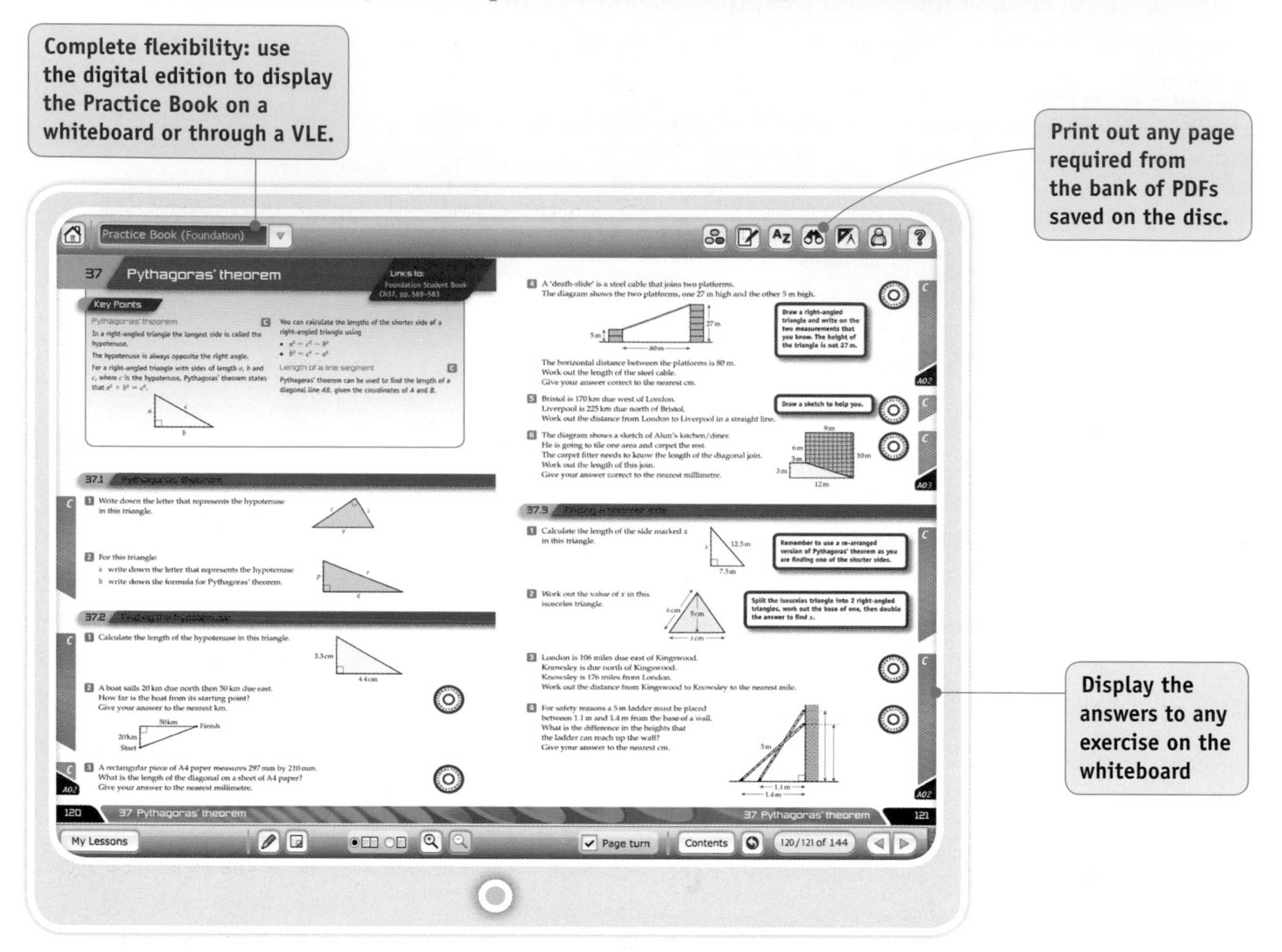

Foundation Tier Resources in the AQA GCSE Mathematics 2010 Series

STUDENT BOOK	PRACTICE BOOK	SUPPORT PRACTICE BOOK	TEACHER GUIDE with EDITABLE CD-ROM
G-C 9781408232750	G-C 9781408232736	G-F 9781408240908	G-C 9781408232743
ACTIVETEACH CD-ROM	**PRACTICE BOOK - Digital Edition**	**SUPPORT PRACTICE BOOK - Digital Edition**	**ASSESSMENT PACK with EDITABLE CD-ROM - Covering all sets**
G-C 9781408232729	G-C 9781408243817	G-F 9781408243824	G-A* 9781408232842

Data collection

Links to:
Foundation Student Book
Ch1, pp. 1–19

Key Points

Data collection F E D

A data collection table or frequency table has three columns: one for listing the items you are going to count, one for tally marks and one to record the frequency of each item.

The data handling cycle D

A statistical investigation follows the data handling cycle:

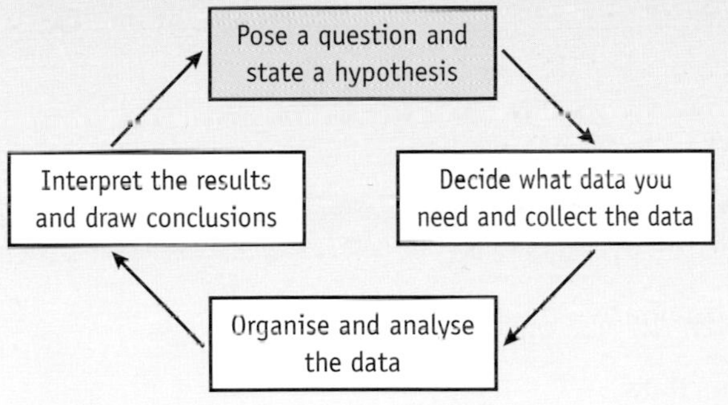

Stating a hypothesis D

A hypothesis is a statement that can be tested to answer a question.

A hypothesis must be written so that it is 'true' or 'false'.

Data sources D

Primary data is data you collect yourself.

Secondary data is data that has already been collected by someone else.

Types of data D

Qualitative data can only be described in words.

Quantitative data can be given numerical values and is either discrete or continuous.

Discrete data can only have certain values.

Continuous data can take any value in a range and can be measured.

Grouped frequency tables for discrete data D

Discrete data can be grouped into class intervals.

Class intervals should be equal sizes.

Grouped frequency tables for continuous data D

To group continuous data, the class intervals must use inequality symbols $\geqslant$ and $\leqslant$.

160 cm $\leqslant h <$ 170 cm means a height from 160 cm up to *but not including* a height of 170 cm.

Recording data in a two-way table D

A two-way table helps you to present related data in a way that makes it easy to answer simple questions.

Questionnaires C

One way to collect primary data is to use a questionnaire.

A questionnaire is a form that people fill in.

Sampling techniques C

The total number of people you could ask to take part in a survey is called the population.

The smaller group of people you ask is called a sample.

A sample that is not representative of the population will be biased.

Random sampling allows every member of the population an equal chance of being selected.

1.1 The data handling cycle

D

1 Write a hypothesis to investigate each question.

- **a** Who can hold their breath longer, students in Year 10 or students in Year 11?
- **b** Do students in Year 10 prefer to wear trainers or shoes?
- **c** Do people who listen to the radio usually prefer Radio 1 or Radio 2?

Make sure that you can test your hypothesis.

2 Give at least one reason why each of these is not a good hypothesis.

- **a** Younger people listen to more loud music than older people.
- **b** Girls talk more than boys.
- **c** People who smoke smell funny.

1.2 Gathering information

D

1 Explain the difference between primary and secondary data.

2 Read these four hypotheses.

a The Arctic has less rainfall per year than London.

b Most of the swimmers at the local swimming pool are boys.

c More 20-year-old girls than 20-year-old boys can cook a curry.

d Girls in my school are better at estimating a length of 50 cm than the boys in my school.

For each hypothesis, state:

i whether you need primary or secondary data

ii how you would find or collect the data

iii how you would use the data.

1.3 Types of data

D

1 Explain the difference between quantitative data and qualitative data.

2 Explain the difference between discrete and continuous data.

3 For each part of the question, write whether the data is quantitative or qualitative data. If it is quantitative data, write whether it is discrete or continuous data.

a The names of students' grandmothers.

b The lengths of pens.

c The number of students in a maths class.

d The makes of calculators used by students.

e The heights of school buildings.

f The score obtained when a ten-sided dice is rolled.

1.4 Data collection

F

1 Alvin asked the students in his class what make of scientific calculator they used. He put the results in table.

Make	Tally	Frequency
do not have one	||	
Casio	~~||||~~ ~~||||~~ |||	
Texas		7
Canon		5
other	|||	
	Total	

a Copy and complete the table.

b What was the most common make of scientific calculator?

c How many more students had a Casio than a Canon?

d How many students did Alvin ask?

2 Steffan rolled a four-sided dice.
These are his results.

1 4 1 3 4 4 1 2 3 3 1 4 4 1 2
3 1 3 3 4 2 2 1 3 4 4 2 1 2 1

a Draw a frequency table to show this information.

b Which number did Steffan roll most often?

c How many times did Steffan roll the dice?

E

3 Members of the Stockport MG car club were asked how many MG cars they owned.
These are their answers.

1 3 1 1 2 3 2 1 4 1 7 1 2 3 2 2
2 4 2 1 1 3 2 1 1 2 2 1 1 2 1 4

a Draw a frequency table to show this information.

b What was the most common answer?

c How can you use the frequency table to work out the total number of MG cars owned by the Stockport MG car club members?

d What is the total number of MG cars owned by the Stockport MG car club members?

FUNCTIONAL • FUNCTIONAL •

E

AO2

1.5 Grouped data

1 This grouped frequency table shows the average speeds of cyclists in a time trial for the Red Dragon cycle club.

Speed, s (km/h)	Frequency
$5 \leqslant s < 10$	10
$10 \leqslant s < 15$	25
$15 \leqslant s < 20$	49
$20 \leqslant s < 25$	11
$25 \leqslant s < 30$	4

a How many cyclists took part in the time trial?

b In which class interval are there the most cyclists?

c How many cyclists had an average speed of less than 25 km/h?

d How many cyclists had an average speed of more than 20 km/h?

e Imagine all the cyclists lined up with the fastest at the front and the slowest at the back. Which class interval would the middle cyclist be in?

FUNCTIONAL • FUNCTIONAL •

D

2 A plant nursery is carrying out research into the height of a new type of foxglove. Fifty seedlings were chosen at random, grown into mature plants and then measured.
These are their heights in cm.

108 111 119 107 104 101 116 102 111 63
118 91 82 105 72 92 103 95 73 110
111 98 99 76 86 96 76 67 94 113
70 117 118 103 65 109 87 99 112 112
113 105 81 90 76 88 66 98 89 103

Design a grouped frequency table to illustrate this data.
Choose suitable class intervals.

AO2

1.6 Two-way tables

D

1 An inspection of students' uniforms and PE kits in Years 7, 8 and 9 gave the following results.

	Correct uniform	Incorrect uniform	Correct PE kit	Incorrect PE kit
Year 7	62	23	54	31
Year 8	81	9	73	17
Year 9	68	18	45	41

a How many Year 7 students had the correct uniform?

b How many Year 9 students had the correct PE kit?

c How many Year 8 students had incorrect uniform?

d How many Year 8 students had incorrect PE kit?

2 A group of 8-year-old children were asked if they could swim or if they were learning to swim. Some of the results are shown in the two-way table.

a Copy and complete the two-way table.

b How many of the 8-year-old children could swim?

c How many of the boys were learning to swim?

	Boys	Girls	Total
Can swim	12	10	
Can't swim		12	20
Learning to swim		7	
Total			54

D

3 Fifty students were asked how they travel to school.
Out of the boys, 15 travel by car, 7 by bus and 5 cycle.
Out of the 22 girls, 10 travel by car, 6 by bus and 4 walk.

a Design a two-way table to show this information.

b Complete the table showing the totals for boys, for girls and for each method of travel.

AO2

c How many boys walk to school?

d How many girls cycle to school?

1.7 Questionnaires

C

1 Calvin wants to find out about what people think about eating 'fast food'.
He includes these questions in his questionnaire.

1) Are you: male ☐ female ☐ ?

2) In what year were you born? ____________

3) Eating 'fast food' is really bad for you.

Do you agree? ________________

4) How often do you eat fast food?

Never ☐ Once a week ☐ Twice a week ☐

3 or 4 times a week ☐ More than 4 times a week ☐

a Is question 1 suitable or unsuitable?

b Is question 2 suitable or unsuitable?

c Explain why question 3 is unsuitable.

d Explain why question 4 is suitable.

2 Dilwyn has seen how poor many children are in Africa.
He wants to find out if students in his school will collect toys and clothes to send to a charity for children in Africa.
Write a suitable question, with a response section, to find out whether other students would collect toys and clothes for the charity.

3 Eleri works for a car insurance company.
She is designing a survey for car owners to complete.
One of her questions is:
'In which month do you renew your car insurance?'

a Design a response section for this question.

b Write a question that she can use to find out the make of car each person owns.

AO2

1.8 Sampling

1 Why is it usual to use a sample when trying to find out information from people living in a town?

2 To find out whether students in a school prefer school dinners or a packed lunch, Nasrin surveyed 10 students in her school canteen.
Give two reasons why her sample is unrepresentative.

3 Marcus carried out a survey into people's opinions on shops being open on a Sunday. He did the survey in front of a supermarket on a Sunday morning.
Is this likely to give a representative sample?
Give reasons for your answers.

AO2

Interpreting and representing data 1

Links to:
Foundation Student Book
Ch2, pp. 20–33

Key Points

Pictograms (G)

A pictogram can be used to represent discrete or categorical data. Each picture represents an item or a number of items. A key is used to show how many items are represented by one picture.

Bar charts (G)

Bar charts and vertical line graphs can show patterns or trends in data.

In a bar chart the bars can be vertical or horizontal. They must be of equal width.

A vertical line graph can also be used to show discrete data.

Comparative bar charts (G F)

You can use dual bar charts to compare data. Two (or more) bars are drawn side by side.

A compound bar chart can also be used to compare two or more sets of data. Two or more bars are drawn on top of each other.

Frequency polygons for grouped data (C)

A frequency polygon shows patterns or trends in the data.

When drawing a frequency polygon for grouped continuous data, the mid-point of each class interval is plotted against the frequency.

2.1 Pictograms

G

1 The pictogram shows the number of Year 10 students who were late to school during one week in May.

a How many Year 10 students were late on Friday?

b How many Year 10 students were late on Wednesday?

c How many more students were late on Tuesday than on Thursday?

AO2

d No one was late on Monday. Give a likely explanation for this.

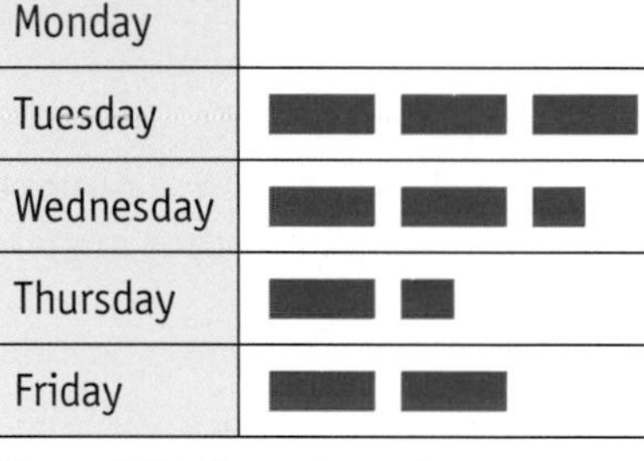

Key: 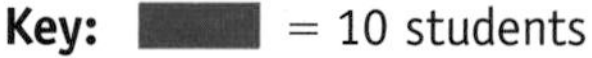 = 10 students

G

2 The pictogram shows the number of cakes sold at a bakery each Saturday in February. No key is given with the pictogram.

a On 10 February, 70 cakes were sold.
How many cakes were sold on 17 February?

b How many more cakes were sold on 3 February than on 17 February?

AO3

c On 24 February, 50 cakes were sold. Copy and complete the pictogram and add a key.

d How many cakes were sold altogether on these Saturdays in February?

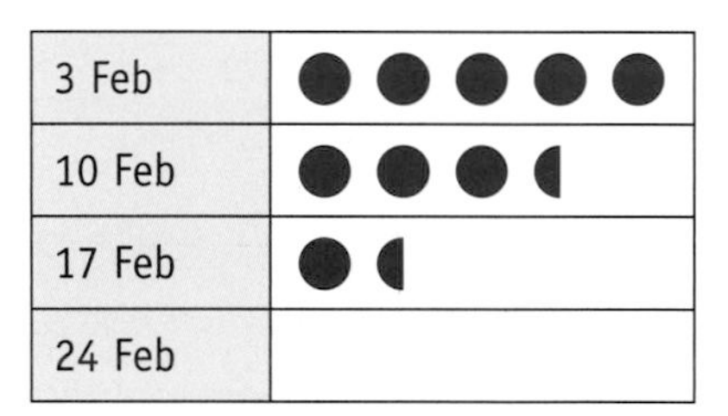

2.2 Bar charts

G

1 The table shows the favourite flavour ice cream of 40 Year 7 students.
Draw a bar chart to represent this information.

Favourite flavour	Frequency
vanilla	22
chocolate	3
strawberry	2
mint	4
caramel	9

2 The bar chart shows the average rainfall in Barcelona and Madrid between September and March.

F

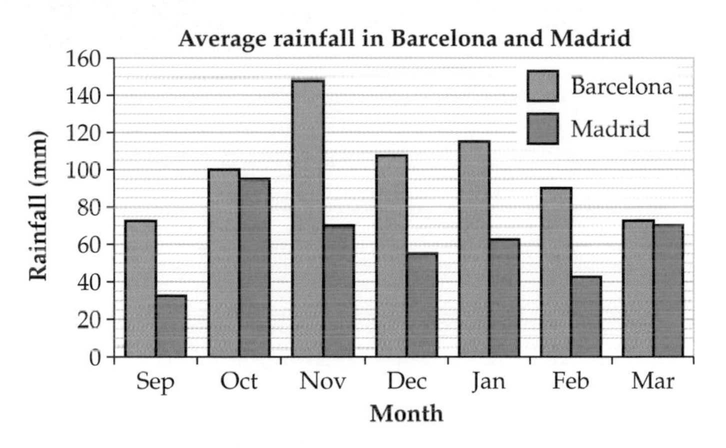

Fay wants to go on a winter city break.
She enjoys walking and sightseeing.

a Use the bar chart above to decide whether Fay should go to Barcelona or Madrid.
Give reasons for your answer.

b What other information should Fay find out to help her make up her mind?

AO2

2.3 Frequency polygons

C

1 The heights of some of the members of the 1992 British Olympic team are shown in the table.

Height, h (cm)	Frequency	Mid-point
$150 \leqslant h < 160$	4	
$160 \leqslant h < 170$	10	
$170 \leqslant h < 180$	14	
$180 \leqslant h < 190$	24	
$190 \leqslant h < 200$	8	

a Copy and complete the table.

b Draw a frequency polygon for this data.

C

2 The frequency polygons show the test results of Mr Andrews' and Mr Stephens' Year 10 maths class.

a Compare the scores of the two classes.
Give reasons for your answers.

b Is it possible to tell from the frequency polygon whether Mr Stephens is a better teacher than Mr Andrews?
Explain your answer.

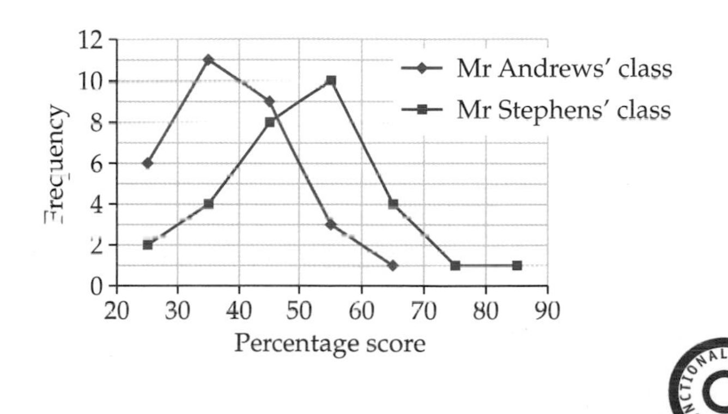

AO3

Links to:
Foundation Student Book
Ch3 pp. 34–58

Key Points

Place value G

The position of a digit in a number tells you its place value.

M	HT	TT	Th	H	T	U
			1	2	3	4

The **2** in 1**2**34 represents **2 hundreds** and has value **200**.

Writing numbers G

A number can be written in words or figures.

Ordering numbers G

To order numbers, start by comparing the digits in the highest place value column.

Rounding whole numbers G

To round to the nearest:

- 10, look at the digit in the units column, for example 2**4** → 20
- 100, look at the digit in the tens column, for example 3**6**2 → 400

If the digit is less than 5, round down. If the digit is 5 or more, round up.

Fraction notation G

A fraction is a part of a whole. The top number is called the numerator. The bottom number is called the denominator.

Equivalent fractions G

To find equivalent fractions, multiply or divide the numerator and denominator by the same number.

Rounding decimals G

To round to one decimal place (1 d.p.), look at the digit in the second decimal place, for example

5.6**3** → 5.6

If the digit is less than 5, round down. If the digit is 5 or more, round up.

Order of operations G

Calculations must be done in the correct order:

Brackets→**I**ndices (powers)→**D**ivision and **M**ultiplication→**A**ddition and **S**ubtraction

Converting metric units F

To convert between different metric units, multiply or divide by 10, 100 or 1000.

Place value in decimals F

A decimal point separates the decimal fractions from the whole number part.

M	HT	TT	Th	H	T	U	.	t	h	th
					5	6	.	7	8	

The **8** in 56.7**8** represents **8 hundredths**

Simplifying fractions F

To simplify a fraction, divide the numerator and denominator by the same number.

Finding a fraction of a quantity F

To find a fraction of a quantity, divide the quantity by the denominator and then multiply by the numerator.

Ordering decimals E

To order numbers, first compare the whole numbers, next compare the tenths, then compare the hundredths, and so on.

Rounding to significant figures E

The first significant figure of a number is the first non-zero digit in the number, counting from the left. To round to one significant figure (1 s.f.), look at the second significant figure. If the digit is less than 5, round down. If the digit is 5 or more, round up.

One quantity as a fraction of another E

To write one quantity as a fraction of another:

- First make sure that both quantities are in the same units.
- Write the first quantity as the numerator and the second quantity as the denominator.
- Simplify the fraction if possible.

The four rules for whole numbers and decimals

When you solve a word problem, look for keywords which tell you which operation to use.

The four rules for fractions G F E D

Make sure you know which is the fraction key on your calculator, and how to use it.

Sensible rounding

When problems involving division do not give whole-number answers, look at the context of the problem to decide whether to round up or down.

3.1 Place value and ordering whole numbers

1 Write these numbers in words.

a 2 b 20 c 200 d 2000

e 20 000 f 200 000 g 2 000 000 h 2 222 222

G

2 Write these numbers in figures:

a four hundred and four

b forty thousand and forty

c four million, four hundred thousand

3 Look at these number cards. 4 2 6

G

a Use all three cards, once each, to make

i the smallest number possible

ii the largest number possible.

b How many numbers can be made with the digit **4** in the hundreds position?

c How many different three-digit numbers can be made using the three cards?
Write all the numbers down.

d Write your numbers from part c in order of size, smallest first.

AO2

3.2 Place value and ordering decimals

1 Write these decimal numbers in words.

F

a 0.2 b 2.22 c 20.02 d 2.202

2 Write down the value of the 4 in each of these numbers.

a 4.1 b 17.4 c 141

d 1.234 e 0.04

3 Write these numbers in order of size, smallest to largest.
0.555 0.55 0.199 0.505

4 Write true or false for each of these.

a $2.1 > 2.09$ b $17.6 > 17.55$ c $8.01 < 8.1$ d $0.2 < 0.135$

5 Look at these cards. 1 2 3 .

F

Use all four cards, once each, to make

a the largest number possible with one decimal place

b the smallest number possible with two decimal places.

AO2

6 Gina is a veterinary nurse. She is preparing to give a dog an injection.
The volume v of the injection must satisfy:

E

FUNCTIONAL

$$14.4 \text{ ml} \leqslant v < 15 \text{ ml}$$

The volume is measured to one decimal place.
What possible volumes could the injection be?

AO2

3.3 Rounding

G

1 Round 34 549 to

a the nearest 10 b the nearest 100 c the nearest 1000.

G

2 The Azores are a group of islands owned by Portugal.
They have a population of 243 000 to the nearest 1000.
Read the statements below and say which **could** be true and which **must** be false.

FUNCTIONAL

a 242 760 people live on the Azores.

b The Azores have 243 520 residents.

c The lowest population the Azores could have is 242 500.

AO2

d The highest population the Azores could have is 243 500.

F

3 A calculator display shows the number

139.9625113

Round the answer to:

E

a three decimal places b two decimal places c one decimal place

d one significant figure e two significant figures f three significant figures.

F

4 Round each of these numbers to an appropriate degree of accuracy.

a A building is 14.8317 m tall.

b The mass of a rock is 125.0355 kg.

c The distance from my house to school is 3.7108 miles.

AO2

d Water, with sea salt added, boils at 101.482°C.

3.4 Converting metric units

F

1 Copy and complete

a 2 km = ☐ m b 2 cm = ☐ mm c 2 kg = ☐ g

d 2 *cl* = ☐ ml e 2*l* = ☐ ml f 2 g = ☐ mg

2 Copy and complete

a 200 m = ☐ km b 200 mm = ☐ cm c 200 g = ☐ kg

d 200 m*l* = ☐ litres e 200 kg = ☐ tonnes f 200 mg = ☐ g

F

3 A jug holds 1 litre.
A cup holds 150 m*l*.
How many cups can be filled from the jug?

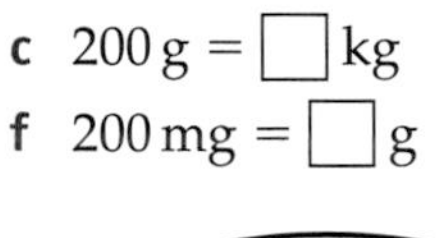

FUNCTIONAL

4 The average weight of a strawberry is 40 g.
A farmer harvests a crop of 300 000 strawberries.
Calculate the mass of the strawberries in

FUNCTIONAL

AO2

a grams b kg c tonnes.

F

5 Write these capacities in order of size, starting with the smallest.

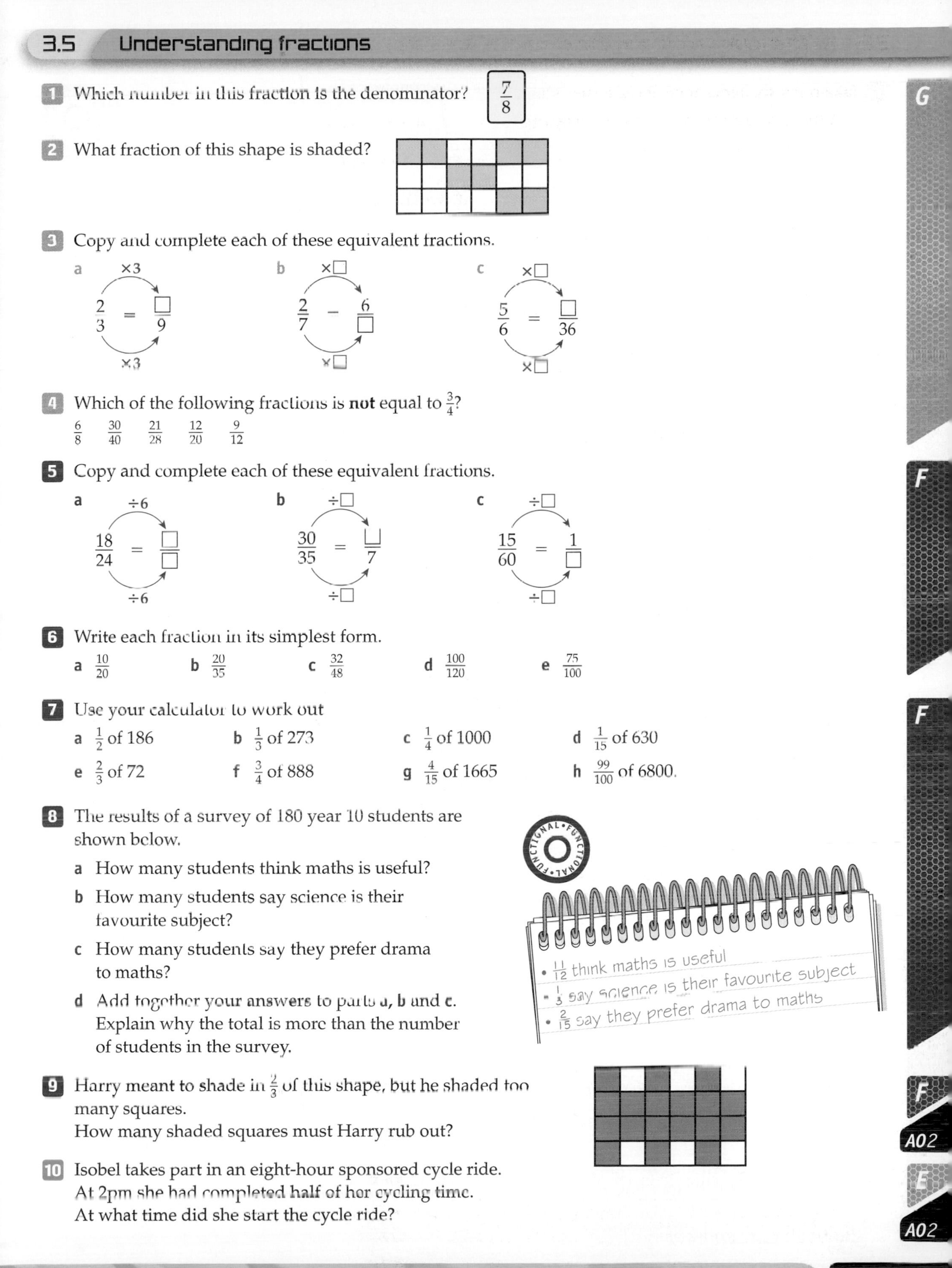

G

1 Which number in this fraction is the denominator? $\frac{7}{8}$

2 What fraction of this shape is shaded?

3 Copy and complete each of these equivalent fractions.

a $\frac{2}{3} = \frac{\square}{9}$ (×3)

b $\frac{2}{7} = \frac{6}{\square}$ (×□)

c $\frac{5}{6} = \frac{\square}{36}$ (×□)

4 Which of the following fractions is **not** equal to $\frac{3}{4}$?

$\frac{6}{8}$ $\frac{30}{40}$ $\frac{21}{28}$ $\frac{12}{20}$ $\frac{9}{12}$

F

5 Copy and complete each of these equivalent fractions.

a $\frac{18}{24} = \frac{\square}{\square}$ (÷6)

b $\frac{30}{35} = \frac{\square}{7}$ (÷□)

c $\frac{15}{60} = \frac{1}{\square}$ (÷□)

6 Write each fraction in its simplest form.

a $\frac{10}{20}$ b $\frac{20}{35}$ c $\frac{32}{48}$ d $\frac{100}{120}$ e $\frac{75}{100}$

F

7 Use your calculator to work out

a $\frac{1}{2}$ of 186 b $\frac{1}{3}$ of 273 c $\frac{1}{4}$ of 1000 d $\frac{1}{15}$ of 630

e $\frac{2}{3}$ of 72 f $\frac{3}{4}$ of 888 g $\frac{4}{15}$ of 1665 h $\frac{99}{100}$ of 6800.

8 The results of a survey of 180 year 10 students are shown below.

a How many students think maths is useful?

b How many students say science is their favourite subject?

c How many students say they prefer drama to maths?

d Add together your answers to parts a, b and c. Explain why the total is more than the number of students in the survey.

F A02

9 Harry meant to shade in $\frac{2}{3}$ of this shape, but he shaded too many squares.
How many shaded squares must Harry rub out?

E A02

10 Isobel takes part in an eight-hour sponsored cycle ride.
At 2pm she had completed half of her cycling time.
At what time did she start the cycle ride?

G

1 Jake buys six items at his local corner shop.

a What is the total cost of Jake's shopping?

b Jake pays with a £10 note.
How much change will he receive?

muffins 2 × 85p
butter 79p
jam £1.29
cheese 2 × £2.40

F

2 Jumana buys three large packets of crisps.
One packet costs £1.49.
In a special offer, three packets cost £3.50.
How much money does Jumana save by buying three packets in the special offer?

3 Copy each calculation.
Put in brackets to make each answer true.

AO2

a $2 \times 6 - 1 = 10$ b $10 \div 3 - 1 = 5$ c $12 \div 6 - 1 = 1$ d $12 - 2 \times 2 = 20$

E

4 Gabi sees these jeans in the window of a shop.

a Work out the reduction in the price of the jeans.

b What is the sale price of the jeans?

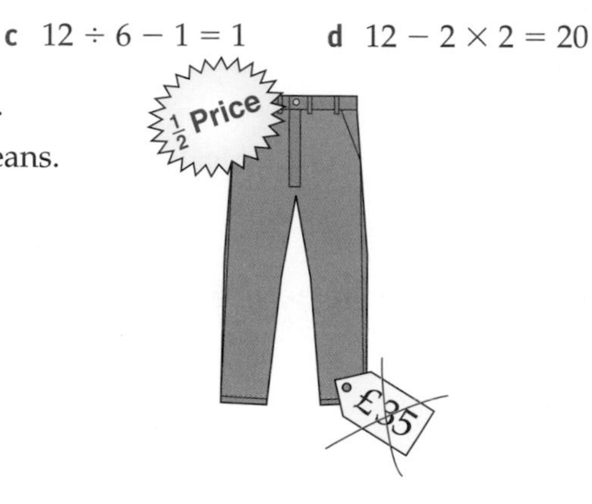

5 Adnan sees this shirt in the window of a clothes shop.

a Work out the reduction in the price of the shirt.

b What is the sale price of the shirt?

SALE $\frac{1}{3}$ off!
£21

AO3

D

6 Krystyna's car needs to be repaired after an accident.
The repair work needed and the costs are shown.

Parts:
1 new tyre @ £72
2 front wings @ £129.95 each
1 bumper @ £107.50
2 front lights @ £38.85 each
2 indicators @ £19.95 each
1 re-spray @ £250
Labour:
6 hours @ £48 per hour

Krystyna could sell the car without doing the repairs for £500.
If she has the repairs done, she could sell the car for £1750.
Is it worth getting the repairs done on the car, or should Krystyna sell the car as it is?
Show all your workings.

AO3

Fractions, decimals, percentages and ratio

Links to:
Foundation Student Book Ch4, pp. 59–74

Key Points

Converting between fractions, decimals and percentages G

To change a percentage into a fraction, write it as a fraction with a denominator of 100.

To change a percentage into a decimal, divide the percentage by 100.

To change a fraction into a decimal, divide the numerator by the denominator.

To change a fraction into a percentage, divide the numerator by the denominator and multiply by 100.

To change a decimal into a fraction, use place value to write it as a fraction with a denominator of 10, 100, 1000 etc.

To change a decimal into a percentage, multiply the decimal by 100.

Calculating the percentage of an amount F E D

Divide the amount by 100, then multiply by the percentage you want to find.

Simplifying ratios E

Divide each of the numbers in the ratio by the highest common factor.

Using ratios E D

Use ratios to compare the lengths of a model to a real-life object.

Writing one quantity as a percentage of another D C

Write the first quantity as a fraction of the second, then multiply the fraction by 100 to convert it to a percentage.

Retail prices index D C

An index number compares one number with another. It is a percentage of the base, and the base is usually 100.

Writing a ratio as a fraction D C

The numerator is the part of the ratio you've been asked about. The denominator is the total number of parts of the ratio.

Writing ratios in the form $1:n$ or $n:1$ C

Look at the number that has to be 1, then divide both sides of the ratio by that number.

4.1 Percentages of amounts

1 Without using a calculator, work out (F)

a 10% of £400 b 5% of £400 c 15% of £400
d 1% of £400 e 11% of £400 f 21% of £400

2 Without using a calculator, work out (E)

a 15% of 6400 km b 12.5% of 6400 km c 0.1% of 6400 km

3 Sterling silver is made from 97.5% pure silver and 2.5% copper. (E, AO2)
A sterling silver teapot weighs 280 g.
What weight of the teapot is copper?

4 Moira bought a painting on eBay for £122.50 and sold it for 46% more than she paid. (D, AO2)
How much did Moira sell the painting for?

5 Ajad compares the prices of the same computer game in two shops. (D, AO3, Functional)
Which shop is cheaper and by how much?

Gareth's Games
Special offer
£36 + VAT (17.5%)

Connie's Consoles
Special deal
£42.50

4.2 One quantity as a percentage of another

D

1 Alf the aardvark found an ants' nest containing 100 000 ants.
He ate 35 000 of them.
What percentage of the ants did Alf eat?

2 Binyamin got £40 for his birthday. He spent £22 on a hoodie.
What percentage of his birthday money did Binyamin spend on a hoodie?

3 Cathy gets 5 weeks holiday a year.
What percentage of the year is Cathy on holiday? Give your answer to two decimal places.

D AO2

4 Last year Darnell got 7 weeks holiday and was also off sick for 2 weeks.
What percentage of last year was Darnell at work? Give your answer to two decimal places.

C

5 Edmundo weighed 88 kg at the start of his diet.
He now weighs 79 kg.
What percentage of his starting weight has he lost? Give your answer to one decimal place.

C AO2

6 Ferdi went shopping and spent £37.50.
She came home with £22.75.
What percentage of her starting amount did she spend?

4.3 Fractions, decimals and percentages

G

1 Write these fractions as decimals.

a $\frac{1}{10}$ b $\frac{2}{10}$ c $\frac{5}{10}$

2 Write these decimals as fractions.

a 0.3 b 0.4 c 0.9

3 Write these percentages as decimals.

a 40% b 50% c 55%

4 Write these fractions as percentages.

a $\frac{1}{2}$ b $\frac{3}{10}$ c $\frac{1}{5}$

5 Write these percentages as fractions.

a 1% b 10% c 25%

6 Write these decimals as percentages.

a 0.8 b 0.35 c 0.07

G AO2

7 Write $\frac{2}{5}$, 0.35 and 38% in order of size, starting with the smallest number.

G AO3

8 Sam wrote $\frac{2}{3} = 66\%$.
Explain why he is wrong.

4.4 Index numbers

D

1 This year the index for a 32″ HD ready TV, compared to 2008 as a base, is 68.

a Has the price of a 32″ HD-ready TV gone up or down?

b By what percentage has the price of a 32″ HD-ready TV changed?

2 In 2006 the average cost of a kilogram of English apples was 90p. Using 2006 as the base year, the price index of apples for the years 2005 to 2009 is given in the table.

Year	2005	2006	2007	2008	2009
Index	94	100	105	102	95
Price (p/kg)		90			

Work out the price of apples from 2005 to 2009.

An index of 94 means that the value has gone down by 6%. An index of 105 means that the value has gone up by 5%.

D

3 The retail prices index was introduced in January 1987.
It was given a base number of 100.
In January 1997 the index number was 154.4.
In January 1987 the 'standard weekly shopping basket' cost £38.50.
How much did the same 'standard weekly shopping basket' cost in January 1997?

AO2

4 The graph shows the exchange rates for the euro and the pound from July 2008 to June 2009.

a What was the exchange rate in November 2008?

b Using July 2008 as the base of 100, work out the index for January 2009.

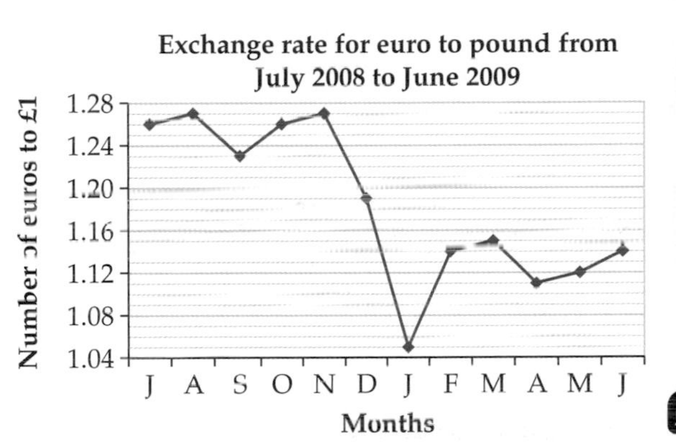

4.5 Ratio

E

1 Simplify the following ratios.

a $5:30$ b $8:12$

c $40:8$ d $2000:150$

Remember to divide both sides of the ratio by the same common factor.

2 Simplify the following ratios.

a 15 m*l* : 50 m*l* b 20p : £1

c 10 seconds : 1 minute d 25 m : 3 km

Remember to make sure both sides of the ratio have the same units.

3 Which of the following ratios is equivalent to $4:5$?

a $25:20$ b $8:10$

c $62.5:50$ d $50:40$

Remember that 1 : 2 is not the same as 2 : 1.

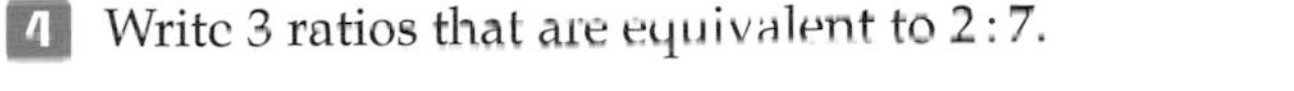

4 Write 3 ratios that are equivalent to $2:7$.

E

5 Jamie makes pastry. He mixes flour and butter in the ratio $3:2$ to make puff pastry.

a Write an equivalent ratio for flour : butter by multiplying both numbers by 5.

Jamie mixes flour and butter in the ratio $5:3$ to make choux pastry.

b Write an equivalent ratio for flour : butter by multiplying both numbers by 3.

c Which type of pastry has the higher proportion of butter?

FUNCTIONAL • FUNCTIONAL •

AO2

E

AO3

6 Zaila makes two different fruit pies.
She uses apples and blackberries in the ratio of 2 : 3.
She uses apples and pears in the ratio of 5 : 8.
Which pie has the higher proportion of apples?
Show workings to support your answer.

FUNCTIONAL

E

7 Huma is making a model of her horse using a scale of 1 : 20.

a Her model is 11 cm long. How long is her horse?

b Her horse is 80 cm wide. How wide is the model?

FUNCTIONAL

8 Neil is building a model battleship using a scale of 1 : 300.

a His model is 80 cm long. How long is the real battleship?

b The real battleship is 33 m wide. How wide is the model?

FUNCTIONAL

E

AO2

9 In Bill's science class there is a model of a very small type of worm.
The scale of the model to the worm is 40 : 1.
The real worm is 0.25 cm long.
How long is the model?

D

10 Aaron finds some information on making concrete.
For every 100 kg of concrete, he needs 14 kg of cement.
How much concrete can Aaron make with:

a 7 kg of cement **b** 28 kg of cement **c** 35 kg of cement?

FUNCTIONAL

D

AO2

11 Here is the list of ingredients that are needed to make a good-quality concrete for a path.
Adele only has 400 kg of sharp sand.
She is using all of the sand to make concrete.
How much of each of the other ingredients should Adele use?

330 kg cement
600 kg sharp sand
1.2 tonnes gravel
180 litres water

FUNCTIONAL

4.6 Ratios and fractions

D

1 The ratio of dogs to cats in a rescue centre is 1 : 2.
There are no other animals at the centre.
What fraction of the animals at the rescue centre are dogs?

2 The ratio of jokes to funny stories in a comedian's show is 5 : 2.
What fraction of the comedian's show is funny stories?

D

AO2

3 A cucumber is made of water and fibre.
The ratio of water to fibre in a cucumber is 19 : 1.
Alun says, '$\frac{1}{19}$ of a cucumber is fibre.'
Is he correct? Explain your answer.

4 $\frac{3}{4}$ of the students in Diane's class have school dinners.
What is the ratio of those who have school dinners to those that don't?

C

5 Lola mixes red and white paint in the ratio 2 : 3 to make pink paint.
She uses 300 m*l* of red paint.
How much white paint does Lola use?

FUNCTIONAL

6 Rachael mixes red, yellow and white paint in the ratio 4 : 5 : 1.
She uses 600 m*l* of red paint.

a How much yellow paint does Rachael use?

b How much white paint does Rachael use?

FUNCTIONAL

C

7 In Ariadne's homework the ratio of wrong answers to correct answers is 1 : 8.
Ariadne got three questions wrong.
How many questions did Ariadne do for homework?

C

AO2

4.7 Ratios in the form 1 : *n* or *n* : 1

1 Write the following ratios in the form 1 : *n*.

a 2 : 7 b 4 : 5 c 10p : £2 d 3 mm : 5 cm

Remember to make sure both sides of the ratio have the same units.

C

2 Write the following ratios in the form *n* : 1.

a 5 : 2 b 150 : 25 c £3.50 : £1.25 d 3 mm : 5 cm

3 Adrian attempted the following question in an exam.

Write 30p : £2 in the form *n* : 1

His answer was 1 : 6.6
Explain the two mistakes that Adrian has made and give the correct answer.

C

4 The ratio of dried fruit to plain flour in Agneta's fruit cake is 9 : 4.

a Write this as a ratio in the form *n* : 1.

The ratio of dried fruit to plain flour in Joe's fruit cake is 68 : 25.

b Write this as a ratio in the form *n* : 1.

c Assuming that Agneta's and Joe's fruit cakes weigh the same, which one has more dried fruit in it?

FUNCTIONAL

AO2

5 Ships' propellers are often made from a mixture of copper, tin and phosphorous.
The ratio of copper : tin : phosphorous in one type of propeller is 450 : 49 : 1.
The ratio of copper : tin: phosphorous in a different type of propeller is 9016 : 961 : 23.
Assuming that the propellers weigh the same, which one has more copper in it?
Show workings to support your answer.

FUNCTIONAL

C

AO3

5 Interpreting and representing data 2

Links to:
Foundation Student Book Ch5, pp. 75–94

Key Points

Interpreting pie charts **F**

A pie chart is a circle that is split up into sectors.

The angles in the sectors add up to 360°. The pie chart must be labelled and the angles accurately drawn.

Drawing pie charts **E**

A pie chart can be drawn from a frequency table.

Stem-and-leaf diagrams **D**

Stem-and-leaf diagrams are used to group and order data according to size, from smallest to largest. A key is needed to explain the numbers in the diagram.

Decimal stem-and-leaf diagrams **D**

Sometimes it may be necessary to have two digits for each of the leaves.

Frequency diagrams for continuous data **D**

Continuous data can be represented by a frequency diagram. A frequency diagram is similar to a bar chart except that it has no gaps between the bars.

Using and plotting scatter diagrams **D**

Scatter diagrams are used to compare two sets of data. They show if there is a connection, called the correlation, between the two quantities plotted.

Lines of best fit and correlation **D C**

On a scatter diagram, a line of best fit is a straight line that passes through the data with an approximately equal number of points on either side of the line.

Positive correlation: as one quantity increases, the other quantity increases.

Negative correlation: as one quantity increases, the other quantity decreases.

No correlation: points are scattered randomly across the diagram.

5.1 Interpreting pie charts

F

1 120 people were asked what type of house they lived in. The pie chart shows the results of the survey.

a Which type of house is the most common?

b What fraction of the people live in a flat?

c How many people live in a flat?

d How many more people live in semi-detached house than live in a detached house?

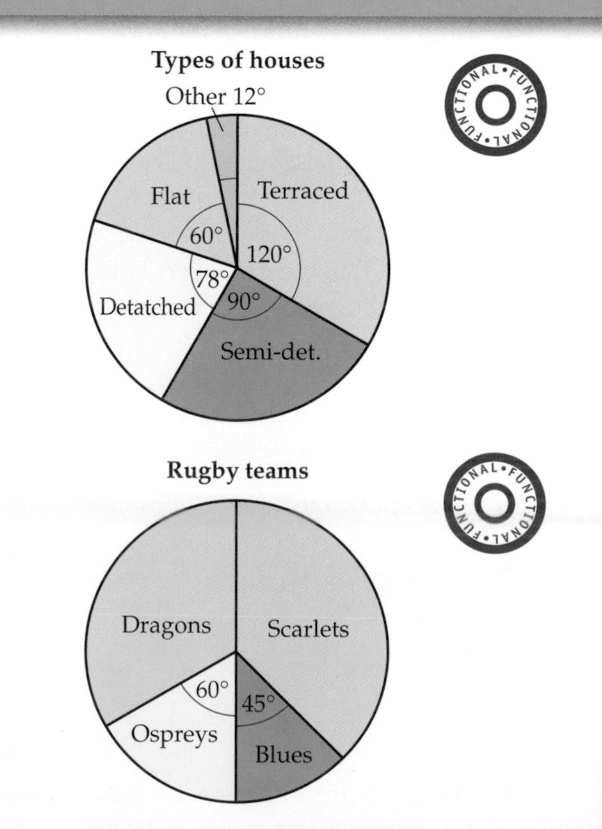

2 The pie chart shows information about the rugby teams supported by some of the students in a Welsh school. Thirty students supported the Blues.

a How many students supported the Scarlets?

b How many students supported the Ospreys?

c How many students took part in the survey?

F

3 Ruth carried out a survey on some Year 7 students.
She asked them what their favourite snack food was.
Abbie carried out the same survey on some Year 8 students.
The pie charts show the results.
In Ruth's survey 40 students chose chocolate.
In Abbie's survey 10 students chose crisps.

Ruth's survey

fruit 150°, crisps 60°, chocolate 120°, sweets 30°

Abbie's survey

fruit 200°, crisps 40°, chocolate 60°, sweets 60°

a How many students in Ruth's survey chose crisps?

b How many students in Abbie's survey chose sweets?

c Ruth says, 'The same number of Year 7 and Year 8 students chose fruit.'
Is Ruth correct? Give a reason for your answer.

FUNCTIONAL

AO2

5.2 Drawing pie charts

E

1 The frequency table shows the sizes of dogs at a boarding kennels one week.
Draw a pie chart to represent this data.

Size of dog	Frequency
small	15
medium	9
large	6

E

2 Lisa carries out a survey on students' favourite athletics event.
Her results are shown in the frequency table.
Lisa decides to draw a pie chart to represent this data.
She starts to work out the angles for each sector.

Athletics event	Frequency	Angle
javelin	12	36°
100 m hurdles	24	50°
100 m	35	
4 × 100 m relay	42	
high jump	7	

a How can you tell that Lisa has already made a mistake in calculating one of the angles?

b Work out the correct angles for this data and draw a pie chart to represent the data.

5.3 Stem-and-leaf diagrams

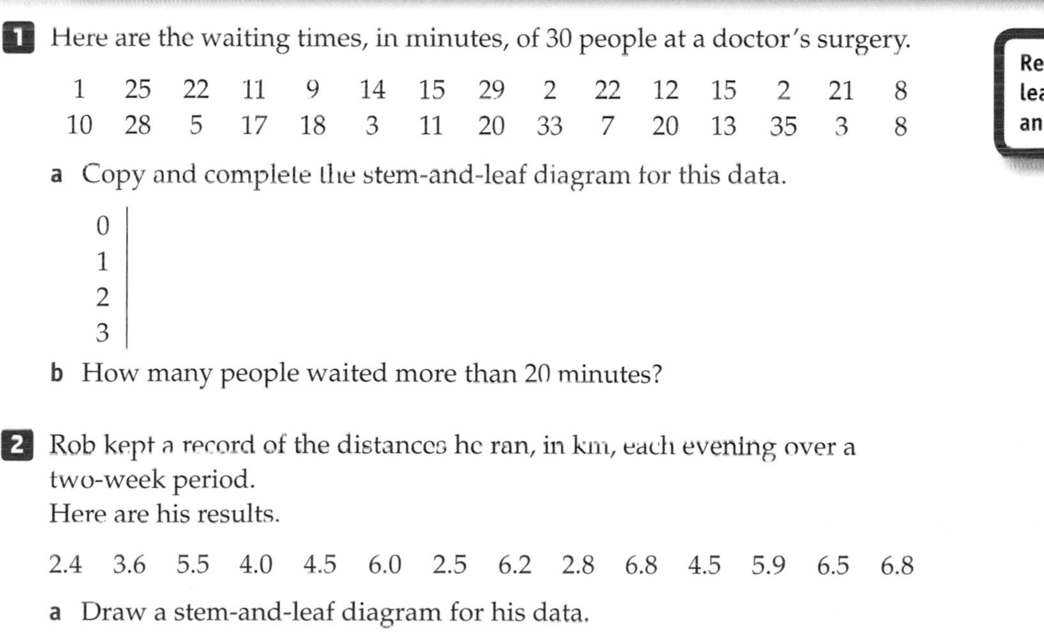

D

1 Here are the waiting times, in minutes, of 30 people at a doctor's surgery.

1 25 22 11 9 14 15 29 2 22 12 15 2 21 8
10 28 5 17 18 3 11 20 33 7 20 13 35 3 8

Remember to put the leaves in order of size and to write a key.

a Copy and complete the stem-and-leaf diagram for this data.

0 |
1 |
2 |
3 |

b How many people waited more than 20 minutes?

2 Rob kept a record of the distances he ran, in km, each evening over a two-week period.
Here are his results.

2.4 3.6 5.5 4.0 4.5 6.0 2.5 6.2 2.8 6.8 4.5 5.9 6.5 6.8

a Draw a stem-and-leaf diagram for his data.

b On how many evenings did Rob run less than 4 km?

D

3 Kiros kept a record of the amount he spent on lunch at work over a four-week period.
Here are his results.

£5.00	£2.58	£5.62	£3.84	£3.65	£4.25	£2.25	£4.06	£3.79	£5.45
£4.12	£3.38	£3.58	£4.85	£2.99	£5.20	£2.80	£3.05	£5.99	£4.66

a Draw a stem-and-leaf diagram for his data.

b What is the difference between the greatest amount and the least amount that Kiros spent on lunch?

c On how many days did Kiros spend more than £4.50?

5.4 Scatter diagrams

D

1 This table shows the marks for 10 students in their maths and music examinations.

Maths	72	35	83	76	59	92	32	28	39	50
Music	64	41	85	59	62	85	24	30	44	46

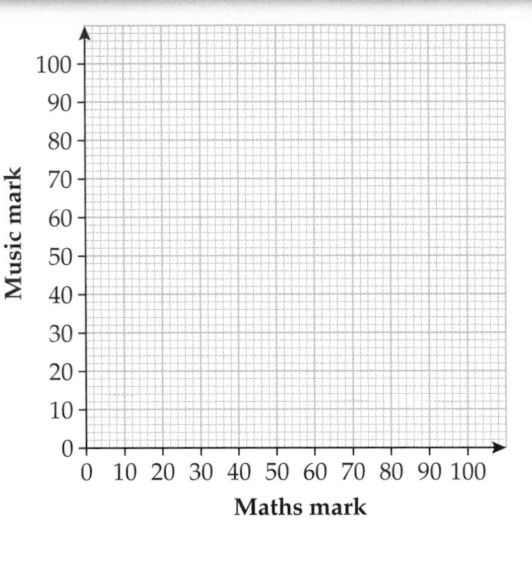

a Plot this information on a scatter diagram.
Use a grid like the one shown.

b Describe the relationship between the maths marks and the music marks of the students.

D

2 Lars is investigating the hypothesis:
'Students who are good at science are not good at foreign languages.'
He plots his data on the scatter diagram shown.

FUNCTIONAL

AO2

a What data did Lars use to test his hypothesis?

b Is Lars' hypothesis correct? Give reasons for your answer.

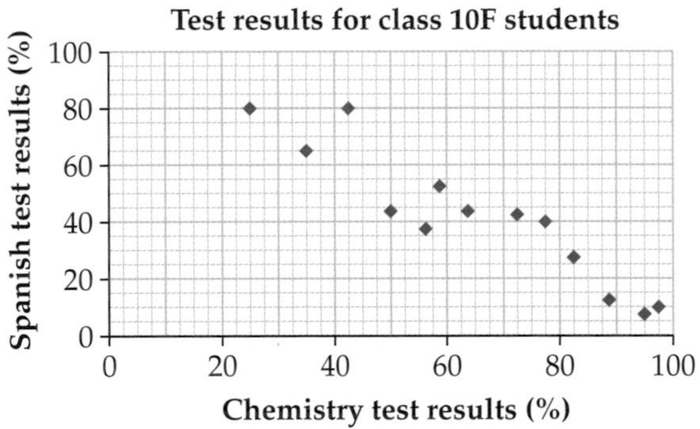

5.5 Lines of best fit and correlation

C

1 Write whether you think each of these statements is likely to be true or false.

a There is a positive correlation between the heights of children and the number of brothers and sisters they have.

b There is no correlation between the age of an adult and the number of the house that they live in.

c There is a negative correlation between the age of a computer and its value.

d There is a positive correlation between the height and weight of a dog.

2 The table shows the average amount spent each week at the supermarket by different families.

Number of people in the family	4	2	3	1	6	1	4	2	4	3
Average amount spent (£/week)	175	90	120	50	250	45	165	85	180	125

a Copy the axes on graph paper and draw a scatter diagram to show the information in the table.

b Describe the correlation between the number of people in the family and the average amount spent each week.

c Draw a line of best fit on your scatter diagram.

d Use your line of best fit to estimate the average amount spent each week by a family of five people.

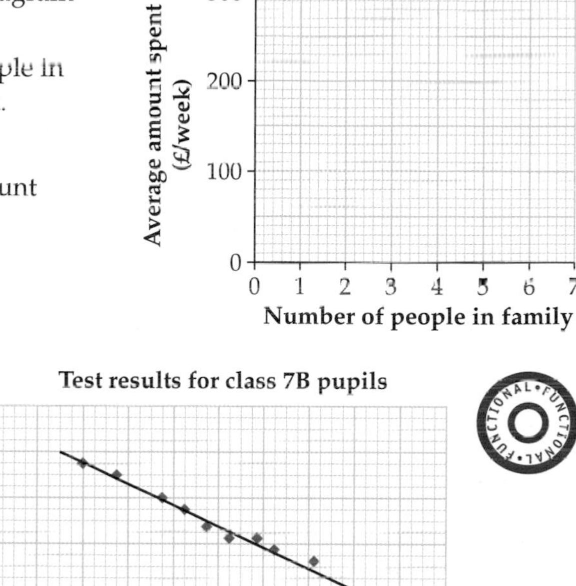

C

AO2

3 The scatter diagram shows the physics and art test results for students in class 7B. A line of best fit has been drawn.

a Describe the correlation between the art results and physics results of class 7B.

b Use the line of best fit to estimate:

i the art result of a student who had a physics result of 65%

ii the physics result of a student who had an art result of 75%.

c Explain why it would not be sensible to use the line of best fit to estimate the art result of a student who had a physics result of 95%.

Test results for class 7B pupils

Physics test results (%): 0, 20, 40, 60, 80, 100

Art test results (%): 0, 20, 40, 60, 80, 100

C

AO3

5.6 Frequency diagrams for continuous data

1 The mass of 200 tomatoes was recorded.
The mass, m, of each tomato was measured to the nearest gram.
The results are shown in the frequency table.
Draw a frequency diagram to show this information.

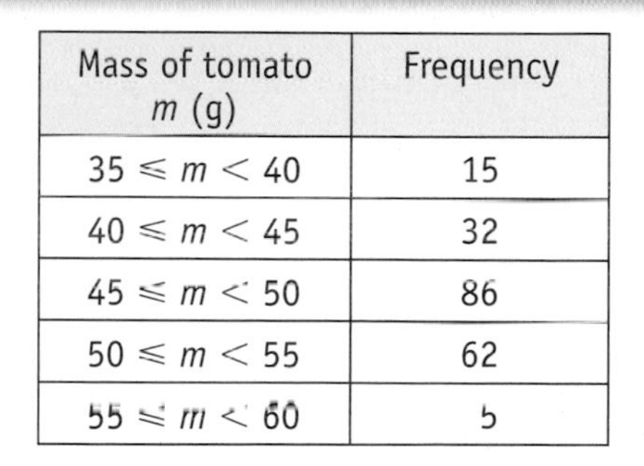

Mass of tomato m (g)	Frequency
$35 \leq m < 40$	15
$40 \leq m < 45$	32
$45 \leq m < 50$	86
$50 \leq m < 55$	62
$55 \leq m < 60$	5

D

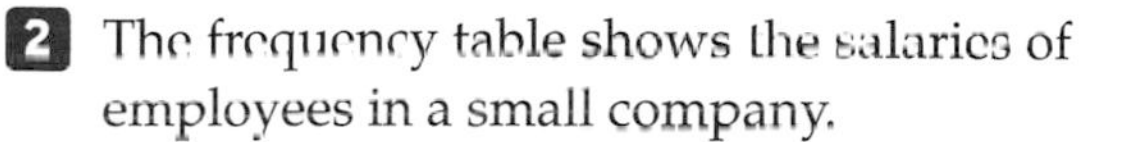

2 The frequency table shows the salaries of employees in a small company.

a Draw a frequency diagram to show this information.

b Fola wants to work towards earning a salary over £40 000. Do you think this is a suitable company for Fola to join?
Give a reason for your answer.

Salary, s (£)	Frequency
$0 \leq s < 10\,000$	2
$10\,000 \leq s < 20\,000$	7
$20\,000 \leq s < 30\,000$	5
$30\,000 \leq s < 40\,000$	3
$40\,000 \leq s < 50\,000$	1

AO2

Range and averages

Links to:
Foundation Student Book
Ch6, pp. 95–110

Key Points

Mode G

The mode, or modal value, of a set of data is the number or item that occurs most often.

Median G F

The median of a set of data is the middle value when the data is arranged in order of size.

Range F

The range of a set of data is the difference between the largest value and smallest value.

The range tells you how spread out the data is.

Mean F E

To find the mean of a set of data, add all the data values together then divide by the number of values.

$$\text{mean} = \frac{\text{sum of all the data values}}{\text{number of data values}}$$

Frequency table averages F E

The range, mode and median can all be calculated when data is presented in a frequency table.

When there are n data values, find the median (middle value) using this formula: median $= \left(\frac{n+1}{2}\right)$th value

Diagram averages G F E

In a pie chart, the mode is the largest sector.

In a bar chart, the mode has the longest bar.

A stem-and-leaf diagram displays data in order of size. This makes it easy to find the range and the median.

6.1 Range

F

1 Tyra kept a record of the amount she spent on lunch during one week at work.

Monday	Tuesday	Wednesday	Thursday	Friday
£3.55	£4.52	£4.80	£3.99	£3.25

Work out the range of these amounts.

2 The maximum daytime temperatures during one week in one town are given below.

12°C 15°C 12°C 17°C 11°C 14°C 14°C

Work out the range for this set of data.

FUNCTIONAL

F

3 Soo recorded the time it took him to complete his homework every evening during one week.
The range of the times was 22 minutes.
The shortest time it took him to complete his homework was 36 minutes.
What was the longest time it took him?

4 John runs a scuba diving club.
During July, John recorded the number of divers that went diving each day.
The highest number of divers that went diving on one day was 16.
The range in the number of divers was 12.
What was the smallest number of divers that went diving on one day?

AO2

F

5 A veterinary nurse weighed a litter of six kittens.
The range in their weights was 30 g.
The nurse wrote the weights of the kittens on a piece of paper, but spilt her tea on the paper.

120 g 105 g 112 g 109 g 125 g

Work out the two possible weights of the sixth kitten.

AO3

6.2 Mode

1 These are the ages of the children in a gym club.

7 8 10 11 12 11 7 9 11 15 13 8 13 11 14

What is the modal value for this data?

G

2 Megan counted the number of sweets in ten different packets. Here are her results.

27 29 28 30 27 25 30 29 27 32

What is the modal value for this data?

3 A mini-market sells eight different bottles of wine.
The prices of the bottles of wine are shown below.

G

£4.99 £5.49 £6.29 £8.99 £5.49 £4.49 £5.49 £6.50

a Work out the modal price of a bottle of wine.

b The manager of the shop increases the price of each bottle of wine by 50p.
Write down the new modal price.

4 Rhys rolled a normal six-sided dice 20 times.
The numbers that he rolled are shown below.

2	3	4	3	5	3	6	3	3	6
1	5	1	5	1	3	6	4	5	3

a Write down the mode for this data.

b Rhys rolled the dice four more times and the mode changed.
What were the four numbers that Rhys rolled?

5 Raisa carried out a survey on the capacities of bottled drinks sold at a supermarket.
Here are her results.

70 *cl* 330 m*l* 1.5 *l* 2 *l* 150 cl 700 m*l* 0.5 *l* 33 cl 250 m*l* 0.7 *l*

Work out the modal capacity.

AO2

6.3 Median

1 The number of minutes that a bus was late at a certain stop was recorded one day.
The results are given below.

G

2 9 8 10 2 5 3 9 7

Work out the median of this data.

2 Paula sells cars. One week she sold five cars.
The value of the cars she sold is shown below.

£5799 £8200 £12 000 £4550 £5899

What is the median value of the cars Paula sold?

3 Danny has five number cards.
All the numbers on the cards are different.
All the numbers are whole numbers less than 10.
The median number is seven.
These are his number cards.

G

Start by writing the numbers in order of size, smallest first.

9	4	1	7	?

What is the missing number on the fifth card?

AO2

F

4 These are Bethan's Year 9 exam marks.

Subject	maths	English	science	French	art	history	music	Welsh
Marks	75%	72%	68%	45%	92%	63%	57%	78%

Work out Bethan's median exam mark.

5 Tao recorded how many text messages he sent each day for 10 days.
Here are his results.

12 23 27 13 8 22 15 19 9 18

Work out the median number of this data.

F AO2

6 In a football team of 11 players, the median age is 23. Each player is a different age.
How many of the players are 23 years old or less?

6.4 Mean

F

1 The number of minutes that trains were late into a certain station were recorded one day.
The results are given below.

12 15 8 4 10 5 12 0 9 17

Calculate the mean of this data.

2 The table shows the number of times Ade rented a DVD each month last year.

Jan	Feb	Mar	Apr	May	Jun	Jul	Aug	Sep	Oct	Nov	Dec
8	7	3	5	4	2	1	6	3	5	6	10

a How many DVDs did Ade rent altogether?

b Calculate the mean number of DVDs that Ade rented each month last year.

F AO2

3 Sara is an estate agent. In one week she sold four houses.
The values of the houses she sold are shown below.

£157 500 £98 500 £112 000 £120 000

a What is the mean value of the houses that Sara sold?

b The local average house price is £125 000.
Is Sara's average house sale price higher or lower than the local average house price?

E AO2

4 The mean of four numbers is 8.5.

a What is the total of the four numbers?

Two of the numbers are 4 and 18. The other two numbers are the same.

b What are the other two numbers?

E AO3

5 Amie has six number cards.
Here are her cards.

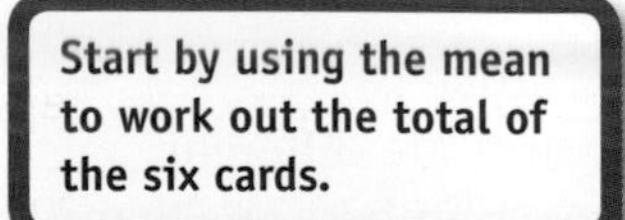

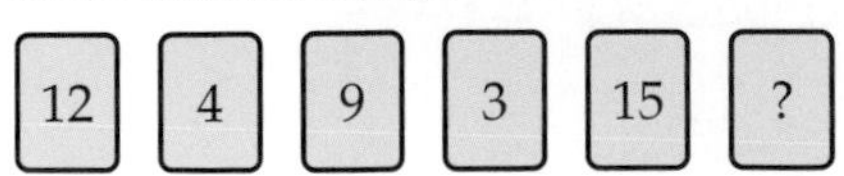

The mean number is eight. What is the missing number on the sixth card?

E AO3

6 Write down three different numbers with a mean of 4.2.

1 A policeman recorded the speeds of 100 lorries, to the nearest 5 mph, on a motorway.
The table shows his results.

Speed (mph)	50	55	60	65	70	75	80
Frequency	9	21	23	24	8	12	3

Work out

a the range of this data

b the modal speed of the lorries

c the median speed of the lorries.

F E

2 Jules carried out a survey on the students in her class.
She asked them how many brothers and sisters they had.
Here are her results.

Number of brothers and sisters	Frequency
0	3
1	16
2	7
3	2
4	1

Work out

a the range of this data

b the modal number of brothers and sisters

c the median number of brothers and sisters.

F E

3 Sheena carried out a survey on the girls in her class.
She asked them what dress size they were.
Here are her results.

Dress size	8	10	12	14	16
Frequency	1	2	4	6	2

Sheena made these three statements:

A 'The range of my data is 5, because $6 - 1 = 5$.'

B 'The mode of my data is 14, because 14 occurred most often.'

C 'The median of my data is 12, because 12 is the middle number.'

a Which of Sheena's statements are correct?

b For Sheena's incorrect statements, explain why she is wrong.

E

4 The table shows the number of computer games owned by the members of a class.

Number of computer games	0	1	2	3	4	5	more than 5
Frequency	7	3	6	8	4	1	2

Is it possible to calculate

a the range **b** the mode **c** the median?

If it is possible, work these out. If it is not possible, explain why.

AO3

G

1 In a survey, students were asked their favourite Olympic sport.
The pie chart shows the results.
What is the mode of this data?

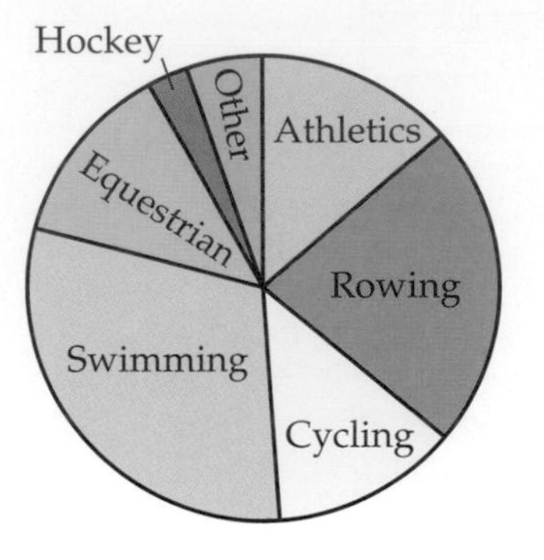

F

2 The bar chart shows the number of goals scored by a hockey team in one season.
Work out

a the modal number of goals
b the range of this data
c the median number of goals.

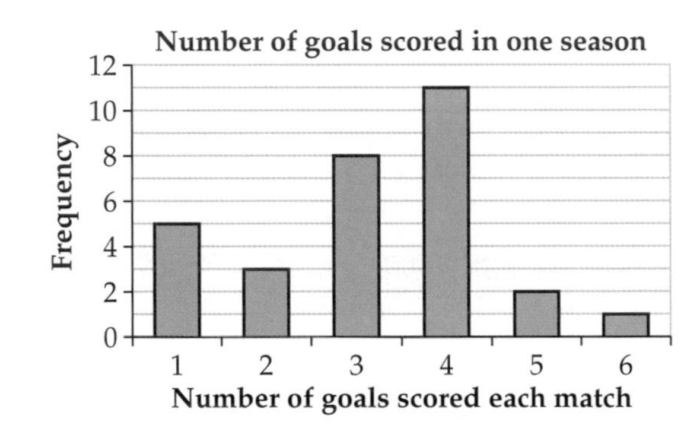

3 The pictogram shows the number of people attending an early morning session at a swimming pool during one week.

a On which day did the most swimmers attend?
b How many swimmers were there on Thursday?
c Work out the range of this data.
d Work out the mean number of swimmers per day.

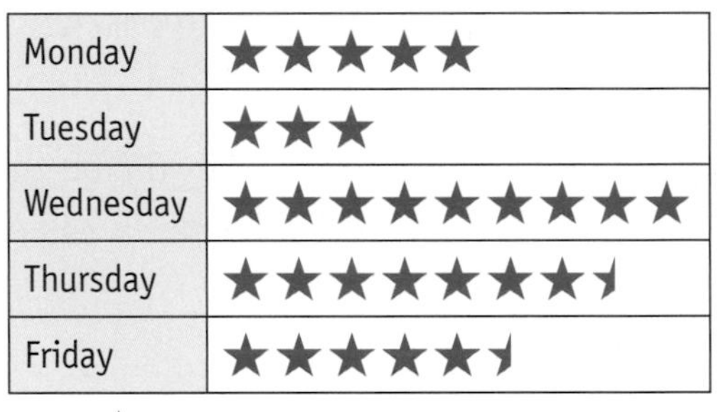

E

4 The stem-and-leaf diagram shows the weights of the cats at a rescue centre.

3	0 1 5 7 9
4	2 3 3 3 5 6
5	3 4 8 9
6	0 1 2

Key: 4 | 2 means 4.2 kg

a Work out the range for this data.
b Work out the median weight.
c Work out the mean weight. Round your answer to one decimal place.

Links to:
Foundation Student Book
Ch7, pp. 115–131

Key Points

What is probability? G

Probability is about measuring the likelihood that something might happen.

The probability scale G

Probability uses numbers and words to describe the chance that an event will happen.

It is measured on a probability scale from 0 to 1.

Writing outcomes G F

To write a list of outcomes, work systematically to make sure that you don't miss any out.

Working out the probability F

The probability of an event happening is

$$\text{Probability} = \frac{\text{number of successful outcomes}}{\text{total number of possible outcomes}}$$

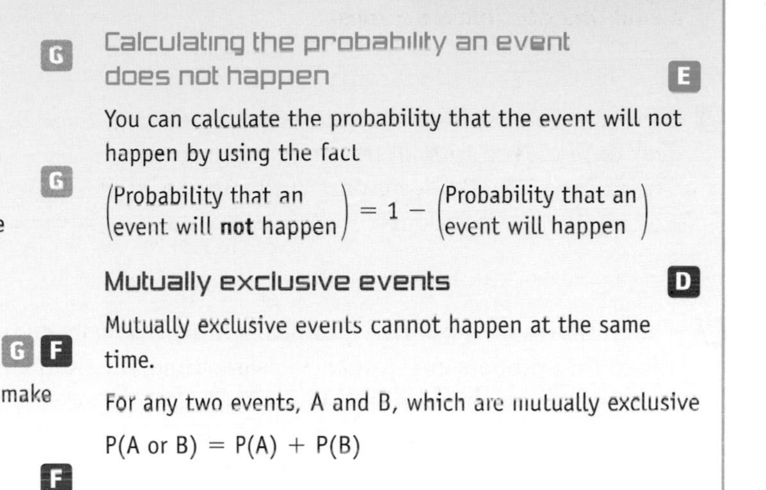

Calculating the probability an event does not happen E

You can calculate the probability that the event will not happen by using the fact

$$\left(\begin{array}{l}\text{Probability that an} \\ \text{event will } \textbf{not} \text{ happen}\end{array}\right) = 1 - \left(\begin{array}{l}\text{Probability that an} \\ \text{event will happen}\end{array}\right)$$

Mutually exclusive events D

Mutually exclusive events cannot happen at the same time.

For any two events, A and B, which are mutually exclusive

P(A or B) = P(A) + P(B)

7.1 The language of probability

G

Write down whether these things are certain to happen, might happen or are impossible.

1. The sun will set on Monday.
2. The day before Thursday is Friday.
3. When you roll a dice you will roll a 2 or a 3.
4. When you roll two dice, both dice will land on 9.
5. When you roll three dice, all three dice will land on 6.
6. You will have a maths lesson on the third of August next year.
7. You will get all of the questions in this practice right.
8. You will breathe out in the next 10 minutes.
9. Next year apples will grow on trees.
10. You will blink soon.

7.2 Outcomes of an experiment

G

1. A normal four-sided dice is rolled. List all the possible outcomes.
2. A coin is taken from a bag containing a 2p coin, a10p coin, a 20p coin and a 50p coin. List all the possible outcomes.
3. This is the list of puddings available in a school canteen. Dylan asked for two puddings. List all the possible combinations of puddings that Dylan could choose.

Today's puddings
Vanilla ice cream
Rice pudding
Chocolate cheese cake
Strawberries

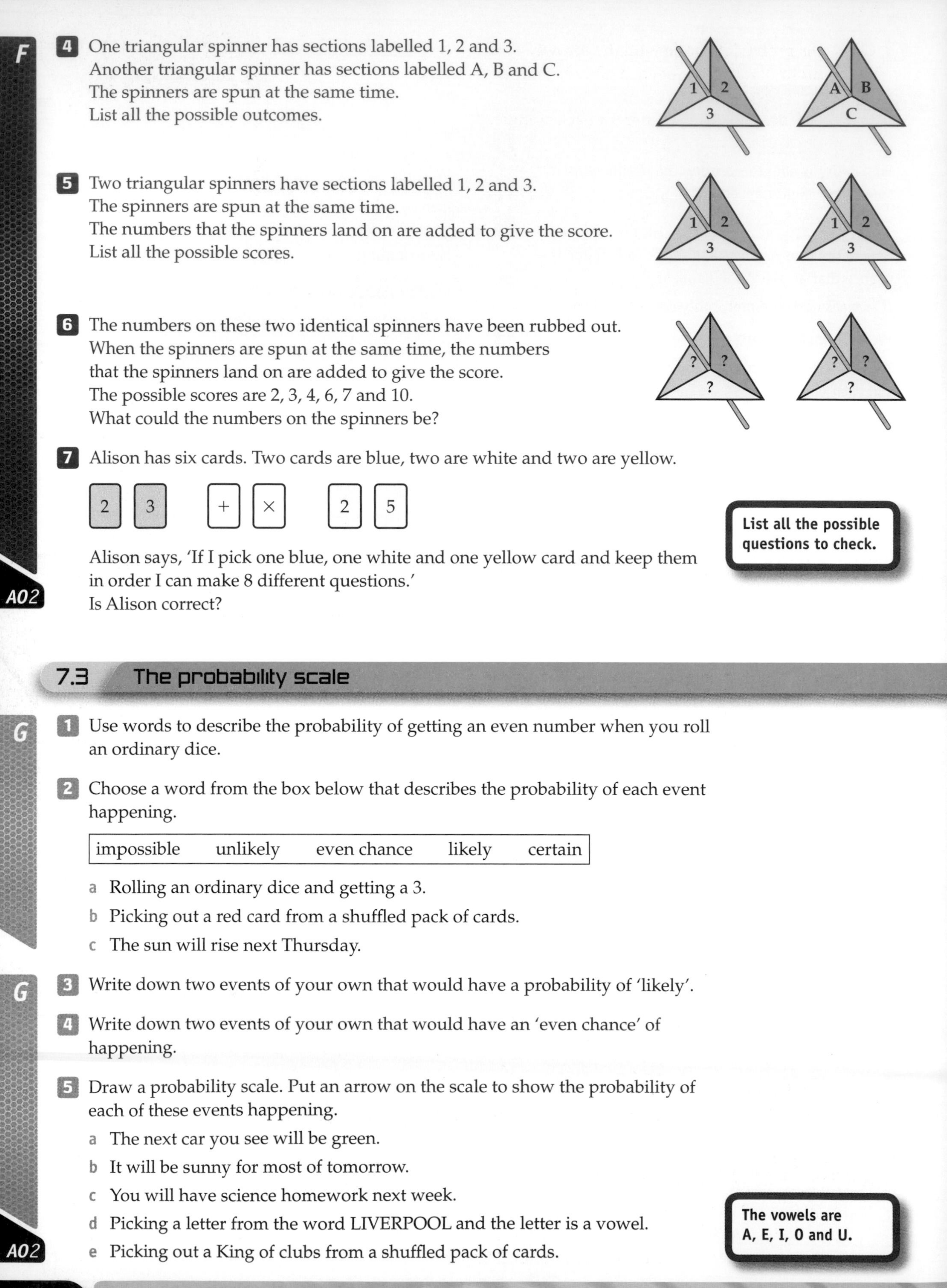

F

4 One triangular spinner has sections labelled 1, 2 and 3.
Another triangular spinner has sections labelled A, B and C.
The spinners are spun at the same time.
List all the possible outcomes.

5 Two triangular spinners have sections labelled 1, 2 and 3.
The spinners are spun at the same time.
The numbers that the spinners land on are added to give the score.
List all the possible scores.

6 The numbers on these two identical spinners have been rubbed out.
When the spinners are spun at the same time, the numbers that the spinners land on are added to give the score.
The possible scores are 2, 3, 4, 6, 7 and 10.
What could the numbers on the spinners be?

7 Alison has six cards. Two cards are blue, two are white and two are yellow.

2 3 + × 2 5

Alison says, 'If I pick one blue, one white and one yellow card and keep them in order I can make 8 different questions.'
Is Alison correct?

List all the possible questions to check.

AO2

7.3 The probability scale

G

1 Use words to describe the probability of getting an even number when you roll an ordinary dice.

2 Choose a word from the box below that describes the probability of each event happening.

impossible	unlikely	even chance	likely	certain

a Rolling an ordinary dice and getting a 3.

b Picking out a red card from a shuffled pack of cards.

c The sun will rise next Thursday.

G

3 Write down two events of your own that would have a probability of 'likely'.

4 Write down two events of your own that would have an 'even chance' of happening.

5 Draw a probability scale. Put an arrow on the scale to show the probability of each of these events happening.

a The next car you see will be green.

b It will be sunny for most of tomorrow.

c You will have science homework next week.

d Picking a letter from the word LIVERPOOL and the letter is a vowel.

e Picking out a King of clubs from a shuffled pack of cards.

The vowels are A, E, I, O and U.

AO2

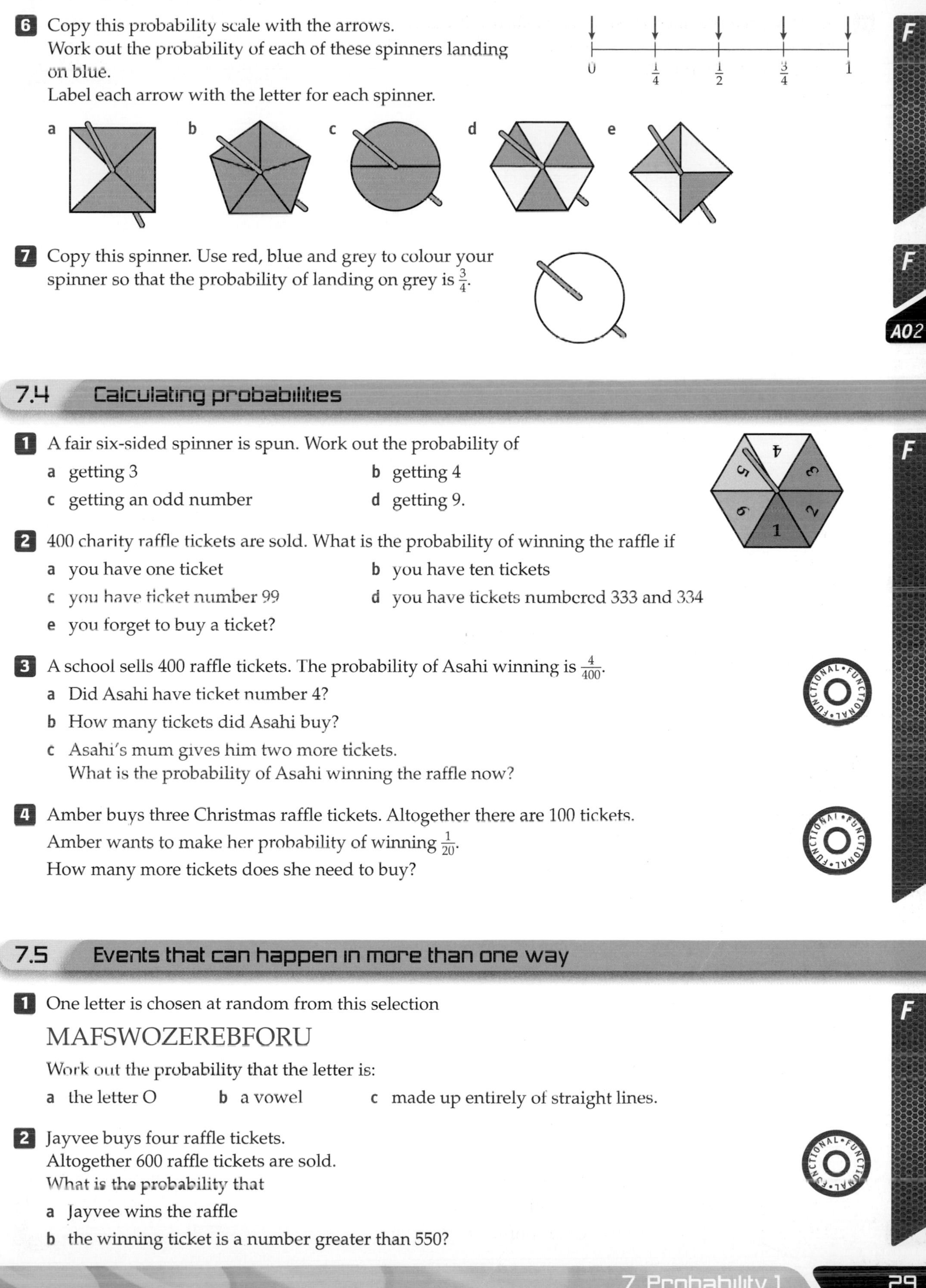

6 Copy this probability scale with the arrows.
Work out the probability of each of these spinners landing on blue.
Label each arrow with the letter for each spinner.

7 Copy this spinner. Use red, blue and grey to colour your spinner so that the probability of landing on grey is $\frac{3}{4}$.

7.4 Calculating probabilities

1 A fair six-sided spinner is spun. Work out the probability of

a getting 3
b getting 4
c getting an odd number
d getting 9.

2 400 charity raffle tickets are sold. What is the probability of winning the raffle if

a you have one ticket
b you have ten tickets
c you have ticket number 99
d you have tickets numbered 333 and 334
e you forget to buy a ticket?

3 A school sells 400 raffle tickets. The probability of Asahi winning is $\frac{4}{400}$.

a Did Asahi have ticket number 4?
b How many tickets did Asahi buy?
c Asahi's mum gives him two more tickets.
What is the probability of Asahi winning the raffle now?

4 Amber buys three Christmas raffle tickets. Altogether there are 100 tickets.
Amber wants to make her probability of winning $\frac{1}{20}$.
How many more tickets does she need to buy?

7.5 Events that can happen in more than one way

1 One letter is chosen at random from this selection

MAFSWOZEREBFORU

Work out the probability that the letter is:

a the letter O
b a vowel
c made up entirely of straight lines.

2 Jayvee buys four raffle tickets.
Altogether 600 raffle tickets are sold.
What is the probability that

a Jayvee wins the raffle
b the winning ticket is a number greater than 550?

3 A card is picked at random from an ordinary pack of playing cards.
What is the probability that the card is

a a King **b** a three **c** a black card **d** a black three?

4 Richard rolls a fair 12-sided dice.
What is the probability that he rolls a number that is

a less than 5 **b** greater than 5 **c** at least 5?

5 Angel has three bags of counters. The bags contain red, blue and green counters.

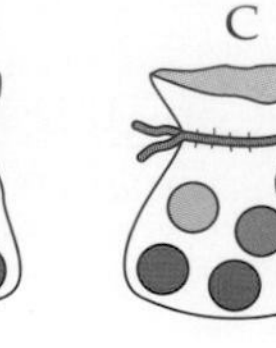

a Which two bags should Angel mix together to give her the highest probability of picking a red counter?

b Angel mixes together the two bags from part **a**.
What is the probability that she picks a blue counter at random from this mixed bag?

6 Adam bought 10 raffle tickets. Hamish said, '500 tickets have been sold altogether.'
Later Adam bought another 10 raffle tickets. Hamish said, 'Now 1000 tickets have been sold altogether.'
Adam said, 'Oh no! I had more chance of winning when I had 10 tickets and only 500 tickets had been sold.'
Explain why Adam is wrong.

7.6 The probability that an event does not happen

1 The probability of picking a toy from a 'lucky dip' box is $\frac{1}{10}$.
What is the probability of **not** picking a toy from the 'lucky dip' box?

2 The probability of picking a spade from a pack of cards is 0.25.
What is the probability of **not** picking a spade from a pack of cards?

3 Gali is learning darts.
The probability that he hits the dartboard is 1 in 3.
What is the probability that his next dart

a hits the dartboard **b** doesn't hit the dartboard?

4 The probability that this spinner lands on 1 is 0.6.
The probability that this spinner lands on blue is 85%.
What is the probability that the spinner

a does not land on 1 **b** does not land on blue?

5 Hannah buys a biased six-sided dice.
The dice is numbered 1 to 6.
The probabilities of different scores are listed in the table.

Number	1	2	3	4	5
Probability	0.15	0.15	0.2	0.3	0.2

Work out the probability of

a not getting a 5 **b** not getting a 2

c not getting a 6. **d** Explain what your answer to part **c** means.

6 SJ has a biased dice numbered 1 to 6.
The probability of getting a 1 with this dice is $\frac{1}{2}$.
SJ says, 'There are 5 other numbers. So the probability of not getting a 1 with this dice is $\frac{5}{6}$'. Explain why SJ is wrong.

7 Leanne has 10 coloured counters in a bag. She picks one at random.
The probability that Leanne **doesn't** pick a blue counter is $\frac{3}{5}$.
How many blue counters does Leanne have in her bag?

7.7 Mutually exclusive events

1 A pencil case contains 20 coloured pencils.
Six of the coloured pencils are red, six are blue, five are green and three are grey.
One coloured pencil is taken from the pencil case at random.
What is the probability that the coloured pencil

a is not grey
b is not red
c is red or blue
d is red or green
e is red or grey
f is not red or grey
g is not green or grey or red?

2 Work out the probability of rolling a 1 or a 2 with a fair dice.

3 A tin contains sweets.
One sweet is taken from the tin at random.
The table shows the probabilities of taking each type of sweet.

a What is the probability that the sweet is a toffee or a chocolate?
b What is the probability that the sweet is a mint?

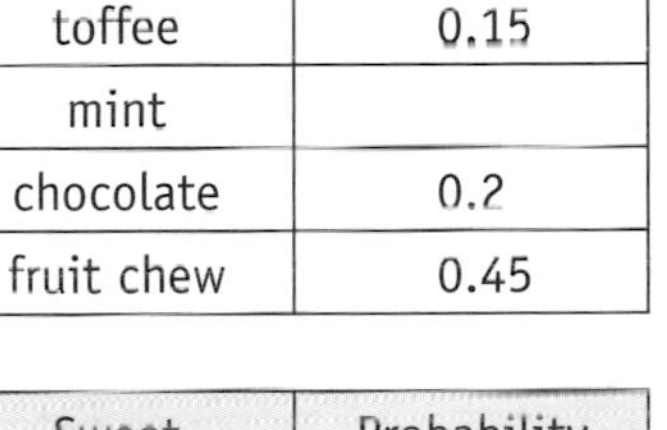

Sweets	Probability
toffee	0.15
mint	
chocolate	0.2
fruit chew	0.45

4 A bag contains sweets.
One sweet is taken from the bag at random.
The table shows the probabilities of taking each type of sweet.
There are five times as many gobstoppers as sherbet bombs.
What is the probability that the sweet is a sherbet bomb?

Sweet	Probability
humbugs	0.2
red rocks	0.2
gobstoppers	
sherbet bomb	

5 Axel puts 12 CDs into a bag.
Bailey puts 8 computer games into the same bag.
Ceri puts some DVDs into the bag.
The probability of taking a DVD from the bag at random is $\frac{1}{5}$.
How many DVDs did Ceri put in the bag?

Links to:
Foundation Student Book
Ch8, pp. 132–147

Key Points

Using frequency diagrams E

Frequency diagrams and frequency tables can be used to calculate the probability of various events happening.

Drawing sample space diagrams E

When two events happen at the same time, all possible outcomes can be shown in a sample space diagram.

This table shows all the possible outcomes when a four-sided dice is rolled and a coin is flipped at the same time.

		Dice			
		1	2	3	4
Coin	H	H1	H2	H3	H4
	T	T1	T2	T3	T4

Probability from two-way tables E

A two-way table shows two or more sets of data at the same time. It shows all the possible outcomes of an event.

The number of times an event is likely to happen D

Sometimes you will want to know the number of times an event is likely to happen.

You can work out an estimate using the formula

Expected frequency = probability of the event happening once × number of trials

Calculating relative frequency C

Relative frequency is also known as experimental or estimated probability.

$$\text{Relative frequency} = \frac{\text{number of successful trials}}{\text{total number of trials}}$$

Independent events C

A and B are independent events if the outcome of one does not affect the outcome of the other.

P(A and B) = P(A) × P(B)

8.1 Frequency diagrams

E

1 The following frequency diagrams and tables show information about the players in a football club.

A Height of players (nearest cm)

13	3 5 5 8
14	0 2 3 4 6 8 9 9
15	0 0 1 1 3 6
16	0 7

Key: 13 | 3 represents 133 cm

B

Weight of players (nearest 5 kg)	
35	☺
40	☺ ◖
45	☺ ☺
50	☺ ☺ ☺ ◖
55	☺ ◖
60	◖

Key: ☺ represents 2 players

C

Number of matches played, m	Frequency
$0 \leqslant m < 10$	3
$10 \leqslant m < 20$	1
$20 \leqslant m < 30$	1
$30 \leqslant m < 40$	5
$40 \leqslant m < 50$	10

A02

D Number of fouls committed

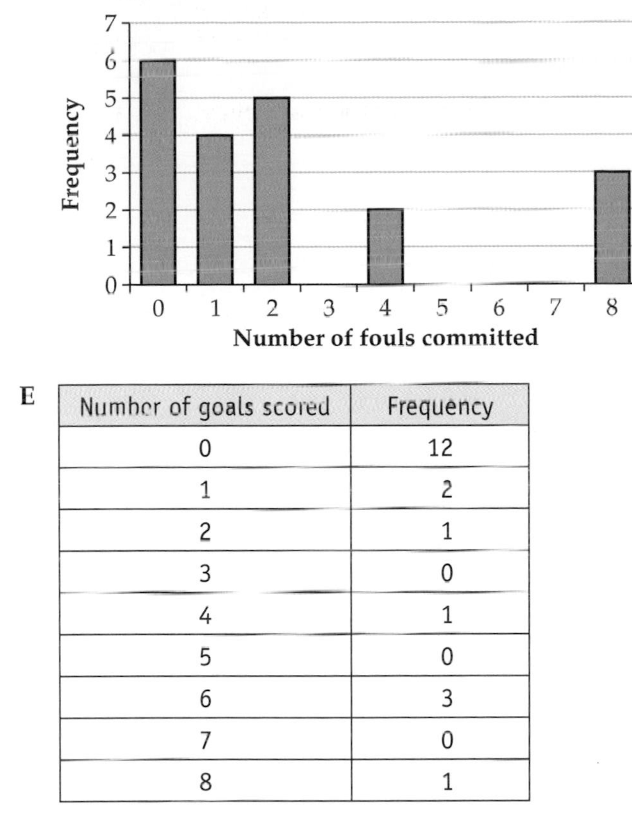

E

Number of goals scored	Frequency
0	12
1	2
2	1
3	0
4	1
5	0
6	3
7	0
8	1

Use the most appropriate diagrams or table to answer the following questions.
A footballer is chosen at random.
What is the probability that the footballer

a has scored four goals
b weighs 55 kg
c is 150 cm tall
d has not scored
e is taller than 160 cm
f has committed no fouls
g has scored more than five goals?

E

AO2

8.2 Two-way tables

E

1 A coin and a four-sided dice are flipped and thrown at the same time.

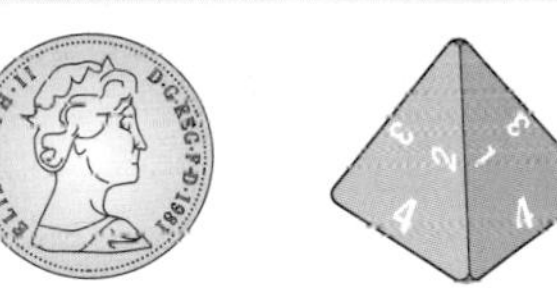

a List all the possible outcomes in a sample space diagram.
b What is the probability of getting a head and a 2?

2 The table shows the number of students in Year 11 who play sport at lunchtime.

	Plays sport	Doesn't play sport	Total
Boys	60	35	95
Girls	45	40	85
Total	105	75	180

One student is chosen at random.
What is the probability that this student

a is a boy who doesn't play a sport at lunchtime
b is a girl who does play a sport at lunchtime
c plays a sport at lunchtime?

E

3 A fair four-sided dice has the numbers 1, 2, 3 and 4.
A different fair four-sided dice has the numbers 1, 3, 5 and 7.
Harry rolls both dice at the same time.
He adds the numbers to give the score.

a What is the probability that Harry gets a score of 6?

b What is the probability that Harry gets a score higher than 6?

AO2

D

4 The tables show the spelling test results for classes 10M and 10G.

10M	Pass	Fail	Total
Boys	11	3	14
Girls	13	3	16
Total	24	6	30

10G	Pass	Fail	Total
Boys	9	3	12
Girls	5	3	8
Total	14	6	20

a A student from 10M is chosen at random.
What is the probability that the student is a girl that has passed the test?

b A student from 10G is chosen at random.
What is the probability that the student is a girl that has passed the test?

c One student is chosen at random.
What is the probability that the student is a boy from 10G that has passed the test?

AO2

8.3 Expectation

D

1 A fair four-sided dice is rolled 800 times.
How many times would you expect it to land on 1?

D

2 The balls from a snooker table are put into a bag.
There are 15 red, 1 yellow, 1 brown, 1 green, 1 blue, 1 pink and 1 black ball.
A ball is taken from the bag at random and then replaced.

a What is the probability that the ball is blue?

b The experiment is repeated 210 times.
How many times would you expect to take a red ball?

3 A fair ten-sided dice has coloured faces.
Four faces are blue, three are yellow, two are red and one is orange.
The dice is rolled 50 times.
How many times would you expect to get

a a blue face

b an orange face

c not a yellow face

d a green face

e a yellow or an orange face

f not a blue or a yellow face?

AO2

D

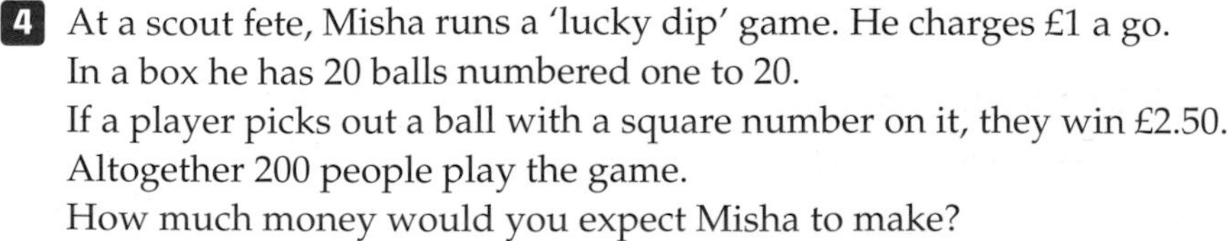

4 At a scout fete, Misha runs a 'lucky dip' game. He charges £1 a go.
In a box he has 20 balls numbered one to 20.
If a player picks out a ball with a square number on it, they win £2.50.
Altogether 200 people play the game.
How much money would you expect Misha to make?

AO3

8.4 Relative frequency

1 In Aberdeen, 120 pedestrians were asked if they had dropped litter in the town in the last month. Seventeen of them answered 'yes'.

a What is the relative frequency of 'yes' answers?

b Approximately how many of the 200 000 people that live in Aberdeen would you expect to have dropped litter last month?

2 Lomo and Mick each carry out an experiment with the same six-sided dice. The tables show their results.

Lomo's results

Number on dice	1	2	3	4	5	6
Frequency	7	2	12	3	5	11

Mick's results

Number on dice	1	2	3	4	5	6
Frequency	30	37	40	31	29	33

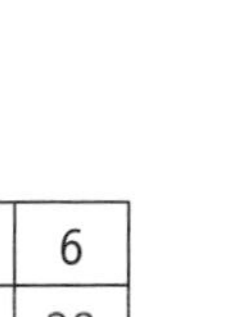

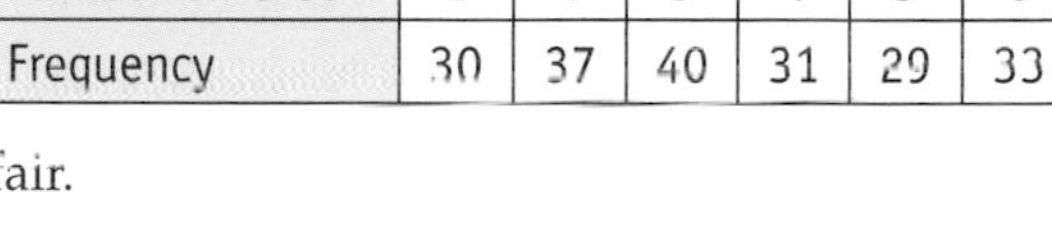

Lomo thinks the dice is biased. Mick thinks the dice is fair.
Who do you think is correct? Explain your answer.

3 Nigel has a four-sided spinner, numbered one to four.
He wants to test his spinner to see if it is biased.
Nigel spins the spinner 20 times. Here are his results.

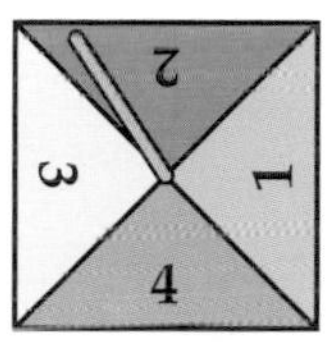

1 2 4 3 2 3 1 2 1 4
2 1 2 3 3 2 1 4 2 3

a Copy and complete the relative frequency table.

Number	1	2	3	4
Relative frequency				

b Nigel thinks that the dice is biased.
Write down the number that you think the dice is biased towards. Explain your answer.

c What could Nigel do to make sure his results are more reliable?

8.5 Independent events

1 A fair four-sided dice is rolled twice.

a What is the probability of rolling a 1 and then a 2?

b What is the probability of rolling two 4s?

2 Otis has 20 pencils in his pencil case. Six of his pencils are HB.
Pete has 12 pencils in his pencil case. Six of his pencils are HB.
Otis and Pete both take a pencil at random from their pencil cases.
What is the probability that they both take a HB pencil?

3 Two fair six-sided dice are rolled together.
The scores on the two dice are added.
What is the probability of getting a total of 2?

4 Quentin rolls a fair four-sided dice.
Rachel rolls a fair six-sided dice.
Work out the probability that

a they both roll a 4

b they roll a total of 2

c they roll a total of 6

d Rachel's score is half of Quentin's score

e they both roll the same number.

Range, averages and conclusions

Links to:
Foundation Student Book
Ch9, pp. 148–162

Key Points

Drawing conclusions

E D

You need to give evidence for your conclusions. This can be information from graphs and tables, or statistics you have calculated like the mean, median, mode and range.

Comparing data

E D

You can use statistics like the mean, median, mode and range to compare two sets of data.

Calculating the mean using an ungrouped frequency table

D

To calculate the mean first work out the sum of all the data values and the total frequency (the number of data values). Then the mean is calculated using the formula

$$\text{Mean} = \frac{\text{sum of all the data values}}{\text{number of data values}}$$

Grouped frequency table averages

The class interval with the highest frequency is called the modal class.

You can estimate the range using the formula

Estimated range = highest value of largest class interval − lowest value of smallest class interval

With n data values, the median is the $\left(\frac{n+1}{2}\right)$th data value.

Population samples

D

When you carry out a survey or experiment you are usually only gathering data from a sample of the population. You can use results from a sample to predict or estimate results about a population, but you can't be certain or exact.

Estimating the mean

C

To estimate the mean first work out the mid-point of each class interval.

$$\text{Mid-point} = \frac{\text{maximum class interval value} + \text{minimum class interval value}}{2}$$

$$\text{Estimate of mean} = \frac{\text{total of 'mid-point} \times \text{frequency' column}}{\text{total frequency}}$$

9.1 Calculating the mean from an ungrouped frequency table

D

1 Rhian counted the number of lorries passing her school gate each minute during one break-time.
Here are her results.

a Copy and complete the table.

b How long was Rhian's break-time?

c How many lorries went past the school gate during Rhian's break.

d Calculate the mean number of lorries per minute passing the school gate at break.

Number of lorries	Frequency	Number of lorries × frequency
0	3	
1	4	
2	6	
3	10	
4	6	
5	1	
Total		

C

2 Steve is collecting information about his school sports day.
The table shows the number of events each student is entered for.

a How many students are in Year 10?

b What is the total number of competitors in all the events in Year 10?

c How many more students entered 4 events in Year 9 than Year 11?

d Calculate the mean number of events entered by Year 9 students.

e Which year group has the highest mean number of events entered per student?

AO2

Number of events	Frequency		
	Year 9	Year 10	Year 11
0	12	17	16
1	29	26	47
2	32	28	31
3	18	17	22
4	21	13	10
5	8	9	4

D

1 Tammy weighed the students in her class.
She recorded her results in this frequency table.

a Write down the modal class.

b Estimate the range of this data.

C

c How many students did Tammy weigh?

d Which class interval contains the median?

Weight, w (kg)	Frequency
$30 \leq w < 35$	3
$35 \leq w < 40$	13
$40 \leq w < 45$	7
$45 \leq w < 50$	3
$50 \leq w < 55$	1

D

FUNCTIONAL

2 In a laboratory, mice are weighed to keep a check on their health.
The results are shown in the frequency table below.

Weight (g)	1–20	21–40	41–60	61–80	81–100	101–120
Frequency	8	14	26	22	8	2

a Write down the modal class.

b Estimate the range for this data.

c Which class interval contains the median?

C

d Amy found that two fat mice, each weighing almost 100 g, had not been included in the frequency table. When she added these two extra pieces of data to the frequency table, what was the effect on

i the modal class

ii the estimated range

iii the class interval containing the median?

AO2

e Tom says that no one could calculate the mean using the data in the frequency table.
Is Tom correct? Explain your answer.

C

3 The grouped frequency table shows the test scores, out of 50, of the students in Miss Read's English class.

Score, s	Frequency	Mid-point	Mid-point × frequency
$0 \leq s < 10$	1	5	$5 \times 1 = 5$
$10 \leq s < 20$	3		
$20 \leq s < 30$	8		
$30 \leq s < 40$	10		
$40 \leq s < 50$	7		
Total			

a Copy and complete the table.

b Work out an estimate for the total test scores of Miss Read's class.

c Work out an estimate for the mean test score of Miss Read's class.

C

4 In an international competition, 60 students from the UK and from Spain sat the same French test. The grouped frequency table shows the scores.
The maximum score was 35.

	Frequency	
Score	UK	Spain
1–5	1	2
6–10	2	5
11–15	4	11
16–20	8	16
21–25	16	10
26–30	19	9
31–35	10	7

a Is it possible to use this table to calculate an exact mean score for each country?
Explain your answer.

b Calculate an estimate of the mean mark for the UK.
Give your answer to one decimal place.

c Calculate an estimate of the mean mark for all 120 students.
Give your answer to one decimal place.

d 'UK students speak better French than Spanish students because the UK students' scores were above average.'
Do you agree or disagree with this statement?
Give reasons for your answer.

AO2

9.3 Drawing conclusions

E

1 Will carried out a survey on some friends. He gave them five different flavour sweets and recorded how many of the flavours they correctly identified.
The table shows his results.

Number of flavours correctly identified	0	1	2	3	4	5
Frequency	9	11	17	21	9	3

Write a conclusion about this data using the mode, the median and the range.

D

2 Xavier and Yvonne carried out a survey on a sample of Year 9 and Year 10 students at their school.
They used different sample sizes. They asked students whether or not they played for one of the school's sports teams.
Here are their results.

Xavier's results	Year 9	Year 10
Do play	32	24
Do not play	19	20

Yvonnes's results	Year 9	Year 10
Do play	9	5
Do not play	4	5

a How many students took part in Xavier's survey?

b A student from Xavier's survey is chosen at random.
Calculate the probability that they play for one of the school's sports teams.

c Calculate the probability that a student from Xavier's survey is in Year 10.

d There are 176 students in Year 10.

i Use Xavier's results to estimate the number of Year 10 students who do not play for one of the school's sports teams.

ii Use Yvonne's results to estimate the number of Year 10 students who do not play for one of the school's sports teams.

e Whose results, Xavier's or Yvonne's, will give the better estimate?
Explain the reasons for your choice.

AO2

E

1 Two golf teams play against each other. There are five members in each team.
In a game of golf, the lower the score the better.
These are the scores of the two golf teams.

Milford team	71	75	69	73	67
Neyland team	65	82	64	71	68

a Work out the mean and range of both sets of scores.

b Write down one similarity and one difference between the scores of both teams.

E

2 Your car has a serious engine problem and needs to be fixed.
You ask two of your local garages how long it has taken them to fix cars with a similar problem in the past. This is the information that they give you.

Greg's garage	15 hours	14 hours	12 hours	17 hours	14 hours
Mick's motors	5 hours	8 hours	23 hours	19 hours	9 hours

a Use the range and mean to compare the times taken by the two garages.

b Both charge the same amount per hour.
Which garage would you choose to fix your car?
Give a reason for your choice.

AO2

D

3 Zaida compared the prices of 2Gb flash drives.
She looked at six different makes on two internet sites.
These are the prices she found.

E-drives	£6.75	£8.12	£7.25	£7.85	£8.95
I-drives	£6.50	£8.50	£6.95	£7.85	£8.75

a Calculate the mean and median of both sets of data.

b Compare the prices of the two internet sites using the mean and the median.

c Overall, which internet site had the cheapest 2 Gb flash drives?
Give a reason for your answer.

D

4 In a survey, some people were asked how many holidays they had taken in Europe in the last 5 years. The results are shown on the bar chart.

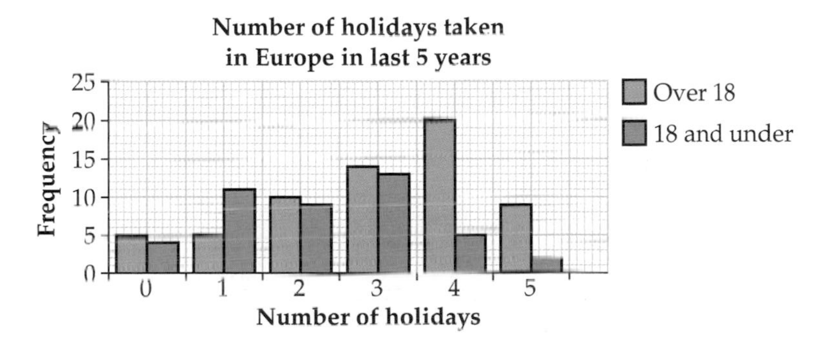

Use the mean, median, mode and range to compare the results for over-18s and 18s-and-under.

AO2

Links to:
Foundation Student Book
Ch10, pp. 163–177

Key Points

Divide a quantity in a given ratio — C

To share in a given ratio:
- Work out the total number of parts to share into.
- Work out the value of one part.
- Work out the value of each share.

Direct proportion — D

When two values are in direct proportion:
- if one value is zero, so is the other
- if one value doubles, so does the other.

Unitary method — D

To solve proportion problems using the unitary method, work out the value of one unit first.

Best buys — D

The best buy means the product that gives you the best value for money.

To compare two prices and sizes, work out the price for one unit.

Exchange rates — D

The exchange rate tells you how many euros, dollars or other currency you can buy for £1. You can use the unitary method to convert between currencies.

Inverse proportion — C

When two values are in inverse proportion, one increases at the same rate that the other decreases.

10.1 Dividing quantities in a given ratio

D

1 Share these amounts in the ratios given.

a £15 in the ratio 1 : 2 b 35 kg in the ratio 4 : 1

C

2 Divide these amounts in the ratios given.

a 45 *ml* in the ratio 2 : 3 b 64 cm in the ratio 3 : 5

3 Share these amounts three ways in the ratios given.

a £84 in the ratio 1 : 2 : 4 b 240 m in the ratio 3 : 5 : 2

C

4 Selia and Vandana buy some raffle tickets between them.
Selia pays £4 and Vandana pays £6 towards the tickets.

a Write the amounts they pay as a ratio.

Selia and Vandana win £2500.

b How much prize money should they each have?

AO2

C

5 A grandmother has three grandchildren, Lisa, Peter and Alex.
She leaves £13 000 in her will to be shared among the three grandchildren in the ratio of their ages.
When their grandmother dies, Lisa is 9 years old, Peter is 6 years old and Alex is 5 years old.

a How much money will Lisa, Peter and Alex receive?

b If their grandmother had lived for two more years before she died, how much money would Lisa, Peter and Alex have received?

AO3

10.2 More ratios

C

1 At a climbing wall the ratio of staff to children climbing is 1 : 3.
How many children can this number of staff take climbing?

a 3 staff b 8 staff

2 During a canoe lesson the ratio of staff to adults going out in a canoe is 1 : 5.
How many staff are needed to take out

a 15 adults **b** 40 adults?

3 Carlos uses 250 g of chorizo sausage and 400 g of chicken in a casserole.

a Write the amounts of chorizo sausage and chicken as a ratio.

b Simplify your ratio.

c How much chicken does he need to go with 200 g of chorizo?

d How much chorizo does he need to go with 1 kg of chicken?

4 In a fruit punch Donna uses 3 litres of orange juice for every 2 litres of pineapple juice.
Donna has 5 litres of pineapple juice.
How much orange juice does she need?

5 The table shows the staff : customer ratio for different ages of customers riding at a stable.
This is the booking information for one ride:

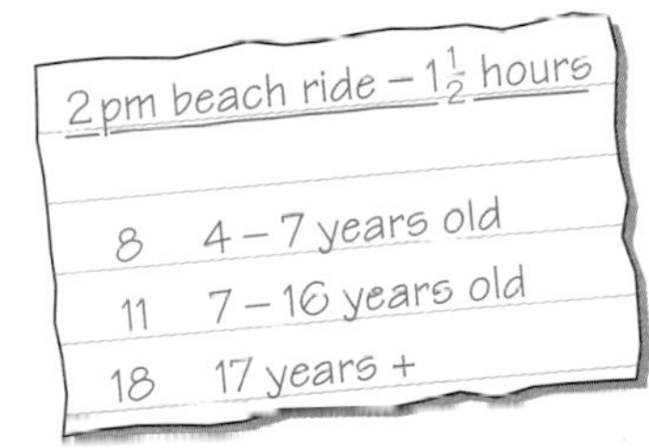

Age of customer	Staff : customer ratio
4–7 years	1 : 1
7–16 years	1 : 3
17 years +	1 : 5

Work out the minimum number of staff needed to take out this ride.

10.3 Proportion

1 Latisha buys 12 tins of dog food for £4.56.

a Work out the cost of 1 tin of dog food. **b** Work out the cost of 5 tins of dog food.

2 Latisha buys 3 packets of dog treats for £7.47.
Work out the cost of 7 packets of dog treats.

3 Latisha takes her dog to training classes.
The costs of the classes are shown on the table.
Latisha takes her dog to the last two weeks of the first course and the whole of the second and third courses.
She only pays for the weeks attended.
How much does it cost Latisha altogether?

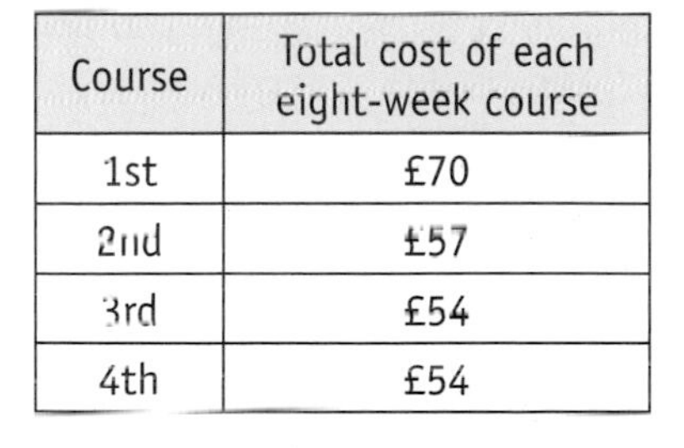

Course	Total cost of each eight-week course
1st	£70
2nd	£57
3rd	£54
4th	£54

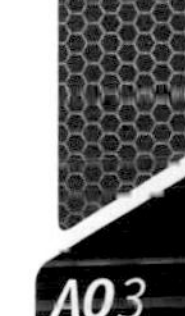

4 Twelve grapefruit cost £4.20.
Work out the cost of

a 24 grapefruit **b** 6 grapefruit **c** 30 grapefruit.

5 Anton gets paid £45 for working 6 hours.
How much does he get paid for working 15 hours?

10.4 Best buys

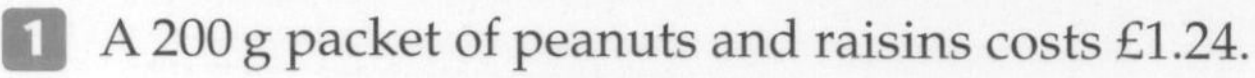

1 A 200 g packet of peanuts and raisins costs £1.24.

a How much does 1 kg cost?

b How many grams of peanuts and raisins do you get for 1p?
Give your answer to two decimal places.

2 A 175 g packet of almonds costs £2.45.

a How much does 1 g cost?

b How many grams of almonds do you get for 1p?
Give your answer to two decimal places.

3 A 240 g pack of cheese costs £1.68. A 400 g pack of cheese costs £2.60.

a Work out the cost of 1 g of cheese in the smaller pack.

b Work out the cost of 1 g of cheese in the larger pack.

c Which pack is the better buy? Explain your answer.

4 A supermarket sells four different boxes of clingfilm.

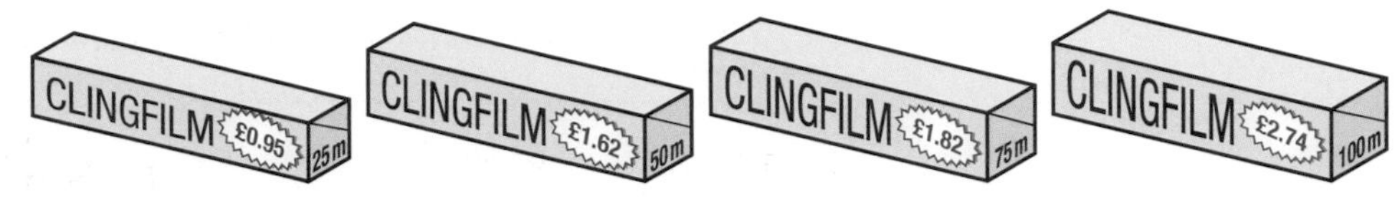

Which box of clingfilm is the best value for money?

10.5 More proportion problems

1 Caroline went on holiday to America when the exchange rate was £1 = $1.80.
She exchanged £350 into American dollars. How many dollars did Caroline receive?

2 Bardo went on holiday to Spain when the exchange rate was £1 = €1.17.
Bardo exchanged £280 into euros. How many euros did he receive?

3 Alicia returned from her holiday in Italy with €80.50.
She exchanged her euros back into pounds when the exchange rate was £1 = €1.15.
How many pounds did she receive?

4 José returned from his holiday in Mexico with 1727 pesos.
He exchanged his pesos back into pounds when the exchange rate was £1 = 22 pesos.
How many pounds did he receive?

5 In September the exchange rate for pounds into Icelandic krona (IK) was £1 = 203.72 IK.
The exchange rate for pounds into Hong Kong dollars (HKD) was £1 = 12.80 HKD.
Anders has 25 465 Icelandic krona. Chen has 1664 Hong Kong dollars.
Who has the most money? You must show all your workings.

6 It takes four people $\frac{1}{2}$ hour to groom and tack up 20 horses.

a How long will it take one person?

b How long will it take six people? Give your answer in minutes.

7 A publisher knows that it takes one proofreader one hour to check two chapters of a book.
The publisher wants the entire book to be checked in six hours.
The book contains 38 chapters.
How many proofreaders does the publisher need to complete the job?

Links to:
Foundation Student Book
Ch11, pp. 184–205

Key Points

Adding and subtracting mentally G

You can use integer complements and partitioning to help you add mentally.

You can find a difference mentally by counting on from the smaller number to the larger number.

Keywords for adding and subtracting G

Look for keywords to help you decide which operations to use to solve a problem.

Multiplying by 10, 100 and 1000 G

To multiply whole numbers:

- by 10, move the digits one place to the left
- by 100, move the digits two places to the left
- by 1000, move the digits three places to the left.

Multiplication facts and partitioning G

You can use multiplication facts and partitioning to help you multiply mentally.

Dividing by 10, 100 and 1000 G

To divide whole numbers:

- by 10, move the digits one place to the right
- by 100, move the digits two places to the right
- by 1000, move the digits three places to the right.

Calculating temperature rise and fall G

Numbers less than zero are negative numbers.

Negative numbers are written with a minus sign (−) in front of the number.

Adding and subtracting whole numbers

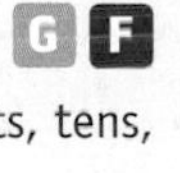

To add or subtract whole numbers, line up the units, tens, hundreds and so on, then add or subtract.

Working backwards G F

You can check the answer to a calculation by working it backwards (using the inverse operation).

Ordering negative numbers

To order negative numbers, think of a number line. Remember that the further to the left you go, the smaller the numbers are.

Adding and subtracting negative numbers F

If you add a negative number, the result is smaller.

If you subtract a negative number, the result is bigger.

Multiplying and dividing negative numbers E

When multiplying or dividing two numbers, you need to check the signs.

- If the signs are the same, the answer is positive.
- If the signs are different, the answer is negative.

Written methods for multiplication G F E

To multiply whole numbers, use the grid method or the standard column method.

Keywords for multiplication include 'multiply', 'times' and 'product'.

Repeated subtraction G F E D

You can use repeated subtraction for division. The number you divide by is called the divisor. Keep subtracting multiples of the divisor until you cannot subtract any more. Then see how many lots of the divisor you subtracted altogether.

Keywords for division include 'divide' and 'share'.

Rounding up and down G F E D

Sometimes the answer to a division is not exact and there is a remainder.

If you are answering a word problem, you may need to decide whether to round up or down.

Checking answers by approximating

You can use estimation to check that an answer is about right.

To estimate:

- round all the numbers to one significant figure
- do the calculation using the rounded numbers.

11.1 Adding and subtracting whole numbers

G

1 Use a mental method to work out

a 35 + 32 b 67 + 23 c 59 − 31 d 55 − 29

2 Copy and complete these additions.

a $6 + \square = 10$ b $25 + \square = 100$ c $23 + 57 + \square = 100$

3 Work out these additions and subtractions.

a $\begin{array}{r} 347 \\ +\ 251 \\ \hline \end{array}$ b $\begin{array}{r} 275 \\ +\ 439 \\ \hline \end{array}$ c $\begin{array}{r} 479 \\ -\ 318 \\ \hline \end{array}$ d $\begin{array}{r} 493 \\ -\ 298 \\ \hline \end{array}$

4 A teacher has a pack of 200 exercise books.
The teacher gives 28 exercise books to one of his classes and 25 to another.
He also gives 135 books to another teacher for her classes.

FUNCTIONAL

a How many books has the teacher given out?

b How many books has he got left in the pack?

5 A souvenir shop has 12 mugs left in stock.
The manager buys another 50 mugs.
During the next week 34 mugs are sold.
How many mugs are left in the shop at the end of the week?

FUNCTIONAL

F

6 Copy and complete.

a $\begin{array}{r} 67 \\ +\ \square\square \\ \hline 89 \end{array}$ b $\begin{array}{r} \square 8 \\ -\ 3\square \\ \hline 12 \end{array}$ c $\begin{array}{r} \square 6 \square \\ +\ 3 \square 7 \\ \hline 891 \end{array}$ d $\begin{array}{r} 4 \square 8 \\ -\ \square 9 \square \\ \hline 135 \end{array}$

7 Portia runs a marathon for charity.
Before the race she raises £7580.
After the race a friend gives her £65 and her brother also gives her some money.
Altogether Portia raises £7725.
How much money did her brother give her?

AO2

11.2 Multiplying whole numbers

G

1 Copy and complete.

a $35 \times 10 = \square$ b $47 \times \square = 47\,000$ c $203 \times \square = 2030$

d $\square \times 100 = 5600$ e $235 \times 100 = \square$ f $\square \times 10 = 445\,600$

F

2 What is the value of the digit 7 in the answer to 273 × 100?

G

3 Use partitioning to work out 43 × 6.

4 Work out these multiplications:

a $\begin{array}{r} 342 \\ \times\ \ \ 2 \\ \hline \end{array}$ b $\begin{array}{r} 431 \\ \times\ \ \ 6 \\ \hline \end{array}$ c $\begin{array}{r} 372 \\ \times\ \ \ 8 \\ \hline \end{array}$

F

5 Work out

a 12 × 36 b 78 × 45 c 246 × 39

6 Saeeda is organising an activity holiday for herself and 11 friends. Return flights cost £135 each. Saeeda has £1800 in her bank account.

a What is the total cost of the flights for the 12 friends?

b Has Saeeda enough money in her bank account to pay for the flights? Show your working.

FUNCTIONAL · E · AO2

7 Berwyn is organising a weekend break for his family. The cost of the weekend break is £268 per person. There are 18 people in his family who want to go on the weekend break. Berwyn has £4200 in his bank account. Is this enough money to pay for the weekend break? Show your working.

FUNCTIONAL · E · AO3

11.3 Dividing whole numbers

1 Copy and complete

a $250 \div 10 = \square$ b $3700 \div \square = 37$ c $4560 \div \square = 456$

d $\square \div 1000 = 6$ e $12\,000 \div 100 = \square$ f $\square \div 10 = 600$

G

2 What is the value of the digit 6 in the answer to $3680 \div 10$?

F

3 Work out

a $141 \div 3$ b $258 \div 6$ c $343 \div 7$

G

4 Four friends go on holiday together.
They travel to the airport by car.
The total cost of petrol and car parking is £92.
They share the cost equally between them.
How much does each of them pay?

FUNCTIONAL

5 Work out these divisions. Some of them have remainders.

a $457 \div 3$ b $326 \div 9$ c $696 \div 12$ d $484 \div 15$

6 A variety pack of crisps contains 15 small bags.

a How many variety packs can be filled from 717 small bags?

b How many small bags are left over?

F

7 Rhys is laying tiles in his kitchen floor. The area of the kitchen floor is 8 m^2.
16 tiles cover an area of 1 m^2. The tiles come in packs of 12.
Each pack costs £25.
What is the total cost of the packs of tiles that Rhys needs to buy?

FUNCTIONAL · E · AO3

11.4 Estimation

1 Use an inverse operation to check these calculations:

a $37 \times 8 = 296$ b $143 + 56 = 199$.

G

2 Use the inverse operation to find the inputs of each function machine.

F

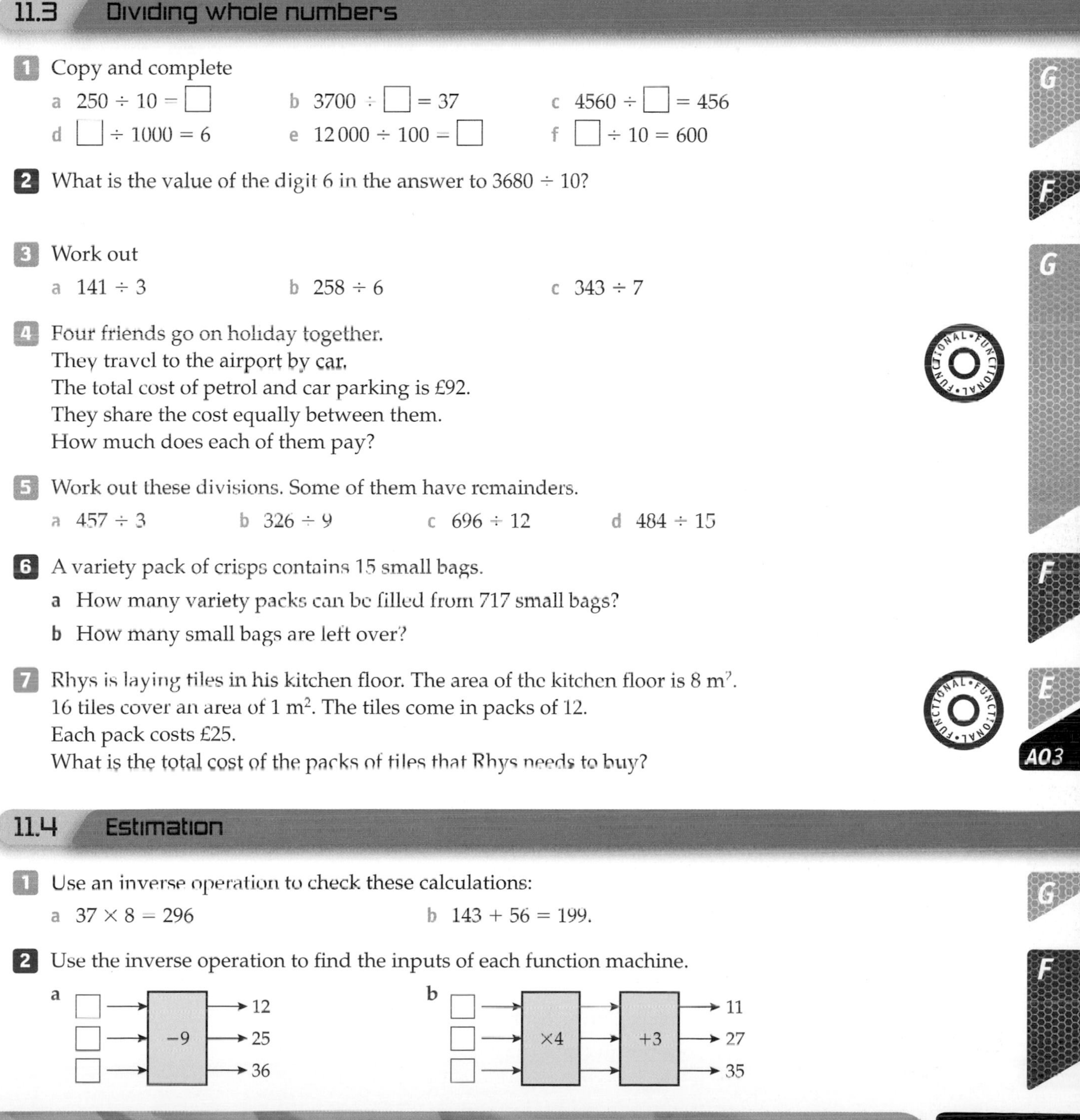

E

3 By rounding to one significant figure, decide which is the best estimate for each question.

a 5.9×34.1	**A** 150	**B** 180	**C** 210
b $585 \div 19.4$	**A** 30	**B** 60	**C** 25
c 47×72	**A** 2800	**B** 4000	**C** 3500
d $831 \div 43$	**A** 18	**B** 20	**C** 16

D

4 Estimate the answer to each calculation. Make sure you write down all your workings.

a $\dfrac{23 \times 4.8}{18.7}$ b $\dfrac{329 \times 586}{32.4}$ c $\dfrac{5.2 \times 512}{71.3 + 28.5}$ d $\dfrac{913 - 288}{37.9 + 11.6}$

C

5 Kathryn is driving to Leeds on the motorway.
She sees this sign.
She is travelling at a speed of 68 miles per hour.
Which of the estimations is the best one to use to work out how long it should take Kathryn to get to Leeds?

Leeds 284 miles

FUNCTIONAL

a $200 \div 70$ b $280 \div 70$ c $300 \div 70$

Give a reason for your answer.

AO2

11.5 Negative numbers

G

1 Write true or false for each of these statements.

a -3°C is warmer than -5°C b 5°C is warmer than -8°C
c -6°C is colder than -10°C d 3°C is colder than -1°C

2 Use the thermometer to help you work out the answers to these questions.

−16 −14 −12 −10 −8 −6 −4 −2 0 2 4 6 8 10 12 °C

Calculate the new temperature if the temperature now is

a 4°C and it cools down by 7°C b -2°C and it cools down by 6°C
c -4°C and it rises by 9°C d -8°C and it rises by 5°C.

F

3 This is part of Payton's homework.

a Which of the questions has Payton answered correctly?
b Which of the questions has Payton got wrong?
c For each question that Payton has got wrong, explain where he has gone wrong and work out the correct answer.

Q1 $5 - (-3) = 5 + 3 = 8$
Q2 $-6 + (-3) = -6 - 3 = -9$
Q3 $2 - (+4) = 2 - 4 = -2$
Q4 $8 - (-5) = 8 - 5 = 3$
Q5 $-12 - (+6) = -12 - 6 = -6$

AO2

E

4 Work out

a $5 \times (-6)$ b $4 \times (-4)$ c $(-3) \times (-7)$ d $(-6) \times (+3)$
e $20 \div (-4)$ f $32 \div (-8)$ g $(-45) \div 9$ h $(-48) \div (-6)$

E

5 a Write down two numbers, one positive and one negative, that have a difference of 5.
b Write down two numbers, both negative, that have a difference of 3.
c Write down two numbers, one positive and one negative, that have a sum of 2.

AO2

Multiples, factors, powers and roots

Links to:
Foundation Student Book Ch12, pp. 208–220

Key Points

Natural and square numbers G F

The numbers 1, 2, 3, 4, 5, … are called natural numbers or counting numbers.

Whole numbers are called integers. Integers can be positive, negative or zero.

Even numbers divide exactly by 2. Odd numbers do not divide exactly by 2.

When a whole number is multiplied by itself, the answer is called a square number.

Finding multiples and factors G F E

When you multiply a number by any whole number, you get a multiple of the first number.

The factors of a number are the numbers that divide into it exactly.

Prime numbers E

A prime number is a number that can only be divided exactly by 1 and itself.

1 is not a prime number. 2 is the only even prime number.

Finding prime factors E

A factor that is also a prime number is called a prime factor.

Finding the least common multiple D C

The least common multiple (LCM) of two numbers is the smallest number that is a multiple of both numbers.

Finding the highest common factor D C

The highest common factor (HCF) of two numbers is the largest number that is a factor of both numbers.

Finding squares, cubes and roots

4×4 is called the square of 4 or 4 squared. It is usually written 4^2.

$4 \times 4 \times 4$ is called the cube of 4 or 4 cubed. It is usually written 4^3.

The inverse of squaring is finding the square root. Use the sign $\sqrt{}$. Positive numbers have two square roots: a positive square root and a negative square root.

The inverse of cubing is finding the cube root. Use the sign $\sqrt[3]{}$.

Using index notation F E D C

You can use index notation to simplify repeated multiplications.

Multiplying and dividing powers C

To multiply powers of the same number, add the indices.

To divide powers of the same number, subtract the indices.

Finding and using prime factors C

To find a number's prime factor decomposition:

- Find any pair of factors of the number. Write them as 'branches' on a tree.
- Keep dividing the factors until you have only prime numbers at the ends of the branches.

Using prime factors to find HCFs and LCMs C

To find the HCF using prime factor decomposition:

- Write each number as the product of its prime factors.
- Find the factors that are common to both lists of prime factors.
- Multiply these to give the HCF.

To find the LCM using prime factor decomposition:

- Write each number as the product of its prime factors.
- For each prime factor, find the higher power in the two lists of prime factors.
- Multiply these to give the LCM.

12.1 Integers, squares and cubes

1 Here are some number cards.
Using the number cards, write down:

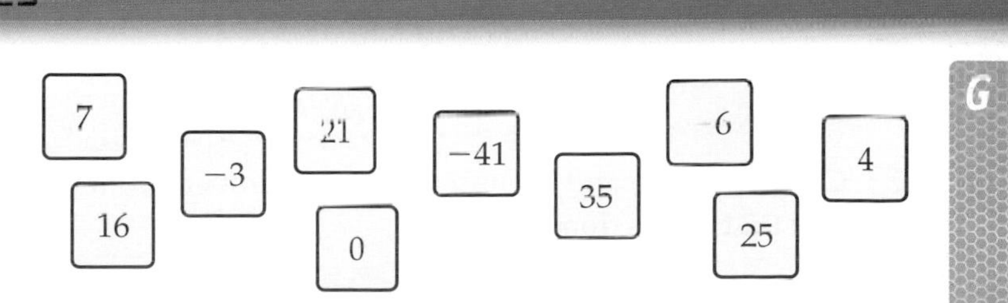

a the largest number
b the smallest number
c the integers between 1 and 10
d the square numbers
e the odd numbers.

G

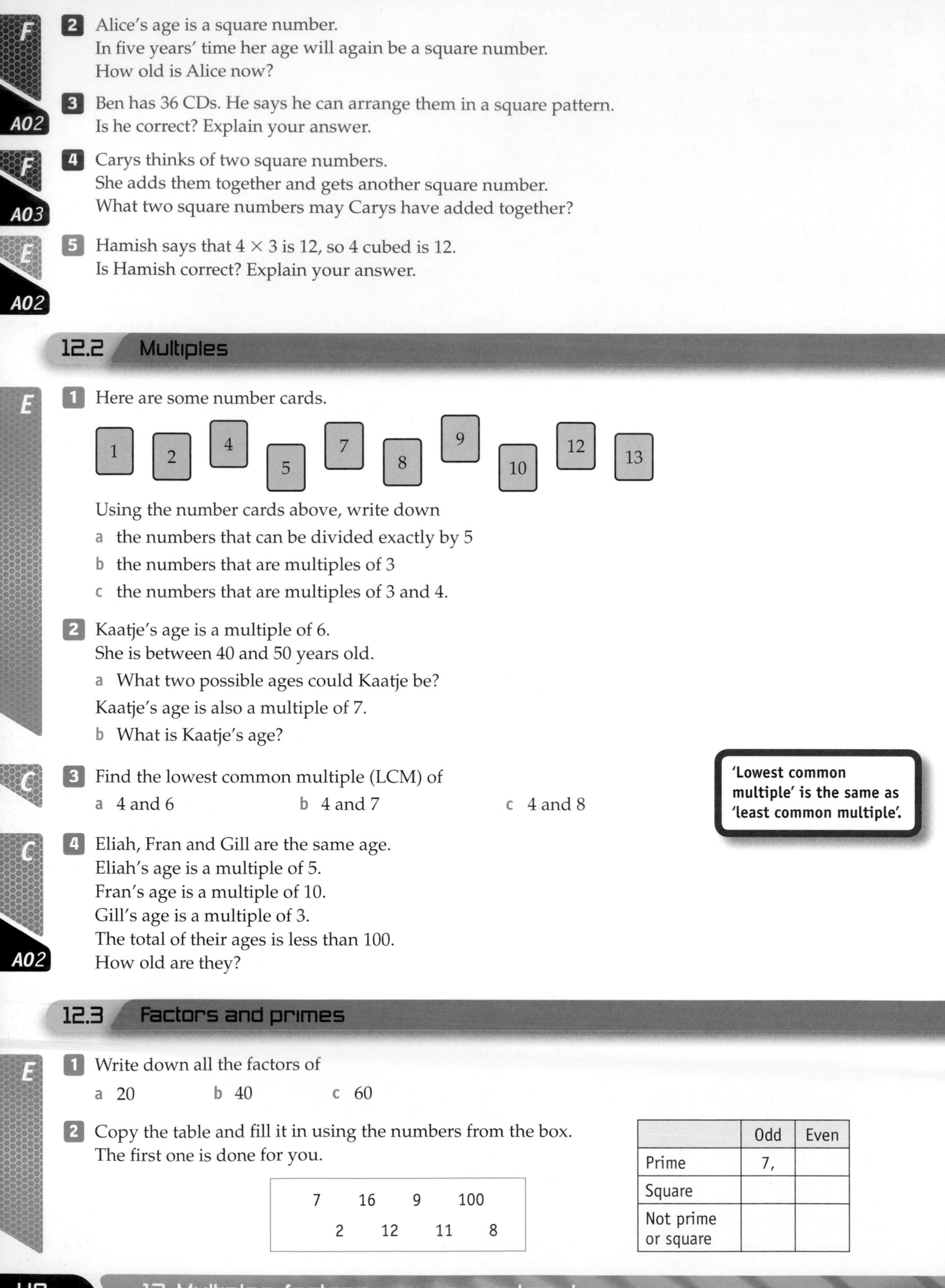

2 Alice's age is a square number.
In five years' time her age will again be a square number.
How old is Alice now?

3 Ben has 36 CDs. He says he can arrange them in a square pattern.
Is he correct? Explain your answer.

4 Carys thinks of two square numbers.
She adds them together and gets another square number.
What two square numbers may Carys have added together?

5 Hamish says that 4×3 is 12, so 4 cubed is 12.
Is Hamish correct? Explain your answer.

12.2 Multiples

1 Here are some number cards.

1 2 4 5 7 8 9 10 12 13

Using the number cards above, write down

a the numbers that can be divided exactly by 5

b the numbers that are multiples of 3

c the numbers that are multiples of 3 and 4.

2 Kaatje's age is a multiple of 6.
She is between 40 and 50 years old.

a What two possible ages could Kaatje be?

Kaatje's age is also a multiple of 7.

b What is Kaatje's age?

3 Find the lowest common multiple (LCM) of

a 4 and 6 b 4 and 7 c 4 and 8

'Lowest common multiple' is the same as 'least common multiple'.

4 Eliah, Fran and Gill are the same age.
Eliah's age is a multiple of 5.
Fran's age is a multiple of 10.
Gill's age is a multiple of 3.
The total of their ages is less than 100.
How old are they?

12.3 Factors and primes

1 Write down all the factors of

a 20 b 40 c 60

2 Copy the table and fill it in using the numbers from the box.
The first one is done for you.

7 16 9 100
2 12 11 8

	Odd	Even
Prime	7,	
Square		
Not prime or square		

3 a Write down all the factors of 50.

b Write down all the prime factors of 50.

E

4 Dory gives two clues about his age:

① My age is a multiple of 4.
② Next year my age will be a prime number.

Explain why you can't work out Dory's age from his clues.

E AO2

5 Find the highest common factor (HCF) of

a 20 and 25 b 20 and 30 c 20 and 32

6 What is the largest number that will divide exactly into 40 and 140?

C

12.4 Squares, cubes and roots

1 Write down the value of

a 5^2 b 10 squared c $\sqrt{144}$

2 Work out

a 5^3 b both square roots of 81 c $3^3 + 2^2$ d $5^3 \div 5^2$

3 Write these numbers in order of size, smallest first.

$\sqrt[3]{1000}$ $\sqrt{225}$ 2^3 3^2

E

4 Write each of these as a fraction in its simplest form:

a $\frac{3^2}{6^2}$ b $\frac{4^2}{6^2}$ c $\frac{3^3}{9^2}$ d $\frac{4^3}{12^2}$

C

5 Hans uses 1 cm^3 blocks to make a cube with side length 5 cm.
He breaks up this cube and makes as many cubes of side length 3 cm as he can.

a How many cubes of side length 3 cm can he make?

b How many 1 cm^3 blocks are left over?

C AO2

12.5 Indices

1 Write these multiplications in index notation.

a 1×1 b $2 \times 2 \times 2$

c $3 \times 3 \times 3 \times 3$ d $100 \times 100 \times 100$

2 Ian says 20^3 is the same as 20×3.
Explain why Ian is wrong.

3 A nursery rhyme starts:
'As I was going to St Ives I met a man with seven wives.
Each wife had seven sacks.
Each sack had seven cats.'
Use index notation to write down how many cats there were.

E

E

4 Work out

a 5^4

b 2^7

c the square of 15

d the cube of 4.

5 a Copy and complete this pattern:

$100^2 = 100 \times 100 = 10000$

$100^3 = 100 \times 100 \times 100 =$ ....

$100^4 =$ .... $=$ ....

$100^5 =$ .... $=$ ....

b Describe what you notice about the pattern.

c Without doing any working, write down the value of 100^8.

12.6 Prime factors

C

1 Write each of these numbers as the product of its prime factors.

a 30

b 14

2 Write each of these numbers as the product of its prime factors.
Write your answer in index form.

a 32

b 108.

3 What number is represented by each of these products of prime factors?

a $2 \times 3 \times 5 \times 7$

b $3^2 \times 5^3$

c $2^3 \times 5 \times 7^2$.

4 Use prime factors to find the LCM of 12 and 42.

5 Use prime factors to find the HCF of 84 and 210.

C AO2

6 A fast food restaurant buys hot dog sausages in packs of 50 and hot dog rolls in packs of 36.
The restaurant wants to buy the same number of sausages as rolls.
What is the **smallest** number of each that they can buy?

FUNCTIONAL • FUNCTIONAL •

12.7 Laws of indices

C

1 Write each expression as a single power.

a $2^3 \times 2^3$

b $3^2 \times 3^6$

c $3^6 \div 3^2$

d $\frac{5^4}{5^2}$

e $\frac{5^2}{5^4}$

2 Write each expression as a single power.

a $4^2 \times 4^3 \times 4^4$

b $5^2 \times 5^2 \times 5^2$

c $9^4 \times 9^5 \times 9$

d $\frac{12^8}{12^4}$

e $\frac{7^3}{7^2}$

f $\frac{8^5}{8^7}$

C AO2

3 Jack says '$2 \times 5 = 10$, so $3^2 \times 3^5 = 3^{10}$.'
Is Jack correct? Explain your answer.

4 Kali says '$2^3 \times 2^2 = 4^5$.'
Explain what Kali has done wrong.

Links to:
Foundation Student Book
Ch13, pp. 229–248

Key Points

Using letters to write simple expressions F

$m + 4$ is called an expression in terms of m.

$3 \times x = x \times 3 = 3$ lots of $x = x + x + x = 3x$

$y \times y = y^2$ ('y squared')

Simplifying algebraic expressions F

To simplify an expression, write it in as short a way as possible.

$a + 2a + 3a$ can be simplified to $6a$.

Collecting like terms F F

You can simplify algebraic expressions by collecting like terms together.

$3a + 6b + 5a = 8a + 6b$

Multiplying terms D

To multiply algebraic terms, multiply the numbers and then multiply the letters.

$3f \times 4g = 3 \times f \times 4 \times g = 3 \times 4 \times f \times g = 12fg$

Expanding brackets D

To expand brackets, multiply each term inside the bracket by the term outside the bracket.

$3(2a + 7) = 6a + 21 \qquad x(3x + 5) = 3x^2 + 5x$

Simplifying expressions with brackets D C

To add or subtract expressions with brackets:

- expand the brackets
- collect like terms to simplify your answer.

Factorising algebraic expressions D C

Factorising an algebraic expression is the opposite of expanding brackets.

- Write a common factor of both terms outside the bracket.
- Work out the terms inside the brackets.

$8t - 10 = 2(4t - 5)$

Expanding two brackets C

To expand two brackets, multiply each term in one bracket by each term in the other bracket.

Grid method

$(x + 2)(x + 5)$

$\times$	x	5
x	x^2	$5x$
2	$2x$	10

FOIL: Firsts, Outers, Inners, Lasts

$(x + 2)(x + 5)$

Firsts: $x \times x = x^2$

Outers: $x \times 5 = 5x$

Inners: $2 \times x = 2x$

Lasts: $2 \times 5 = 10$

13.1 Using letters to write simple expressions

F

1 Use algebra to write expressions for these.

- **a** 4 more than y
- **b** 7 less than x
- **c** b added to 9
- **d** 6 lots of d
- **e** q divided by 2
- **f** 12 multiplied by p

2 Write these expressions in a simpler form.

- **a** $n + n$
- **b** $f + f + f + f$
- **c** $k + k + k$

3 Use algebra to write expressions for these statements.
Use n to represent the unknown starting number.
The first one is done for you.

- **a** I think of a number and add 5. (Answer: $n + 5$)
- **b** I think of a number and subtract 4.
- **c** I think of a number and multiply by 7.
- **d** I think of a number and divide by 8.
- **e** I think of a number, multiply it by 2 then add 9.
- **f** I think of a number, divide it by 2 then subtract 3.

E

4 Trish has 3 packets of biscuits.
Each packet has g biscuits in it.

a How many biscuits does she have altogether?

b Trish eats 4 of the biscuits.
How many biscuits does she have left?

D

5 Write an expression for the total cost, in pence, of

a 3 oranges at x pence each and 5 grapefruit at y pence each

b 2 muesli bars at m pence each and 7 chocolate bars at n pence each.

13.2 Simplifying algebraic expressions

F

1 Simplify these algebraic expressions.

a $3b + b$　b $3t + 8t$　c $5y - 2y$　d $15m - 9m$

e $4p + 3p + 7p$　f $6x + 2x - x$　g $3h - h + 12h$　h $5g + 2g + 6g - 3g$

F A02

2 In an algebraic pyramid the expression in each block is found by adding the two expressions below it.
Copy and complete these pyramids.

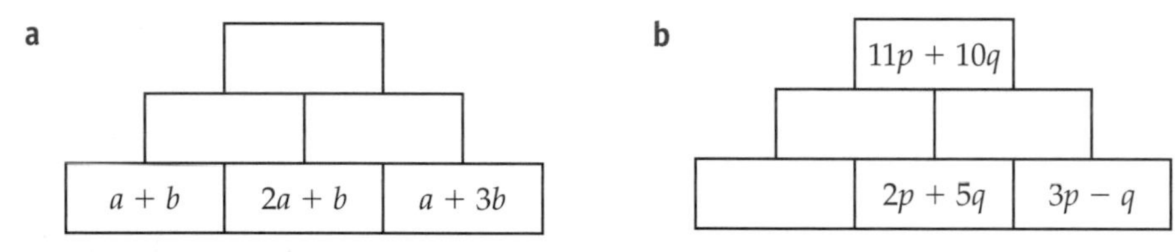

E

3 Simplify these expressions by collecting like terms.

a $3x + 2y + 4x$　b $5t + 4 + 8t + 1$　c $5h + 3h + 2h + 6j$

d $8k - 3k + 5p - 3p$　e $2q + 9n + 3q - 6n$　f $8w + 10 - 3w - 3$

g $4pq + 3pq$　h $3ab + 5ab - ab$　i $4t + 9xy - 2t + 3xy$

j $4x^2 + 9 + 3x^2 - 7$　k $8gh + 4gk - 3gh + 5gk$　l $12d^2 + 20de - 3d^2 - 13de$

E A02

4 Write expressions for the perimeters of these rectangles.
Write each expression in its simplest form.

a

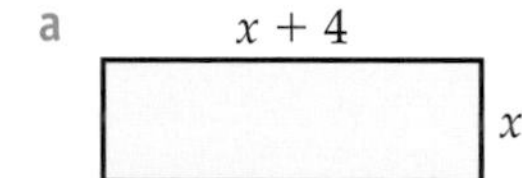

b

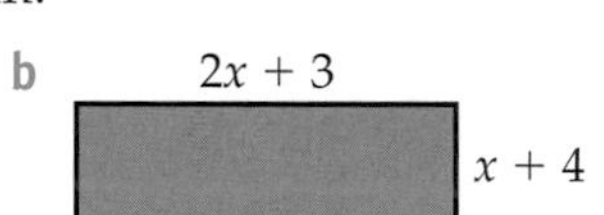

13.3 Multiplying in algebra

E

1 Simplify

a $3 \times 3k$　b $4 \times 6a$　c $8 \times 3m$　d $2p \times 9$

e $7x \times 2y$　f $2m \times 5n$　g $3t \times 4t$　h $4d \times 6d$

E A02

2 In this algebraic pyramid the expression in each block is found by multiplying the two expressions below it.
Copy and complete this pyramid.

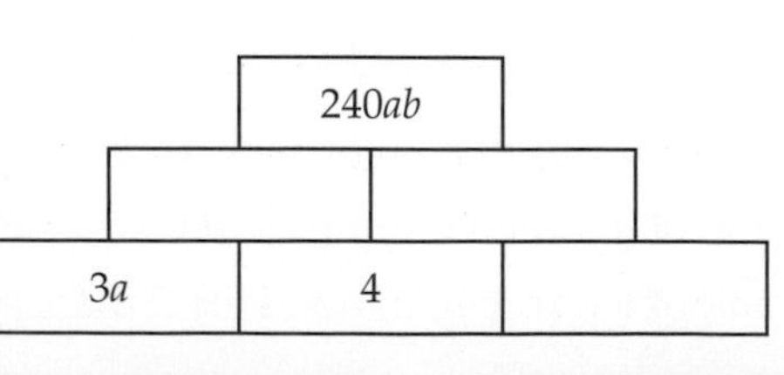

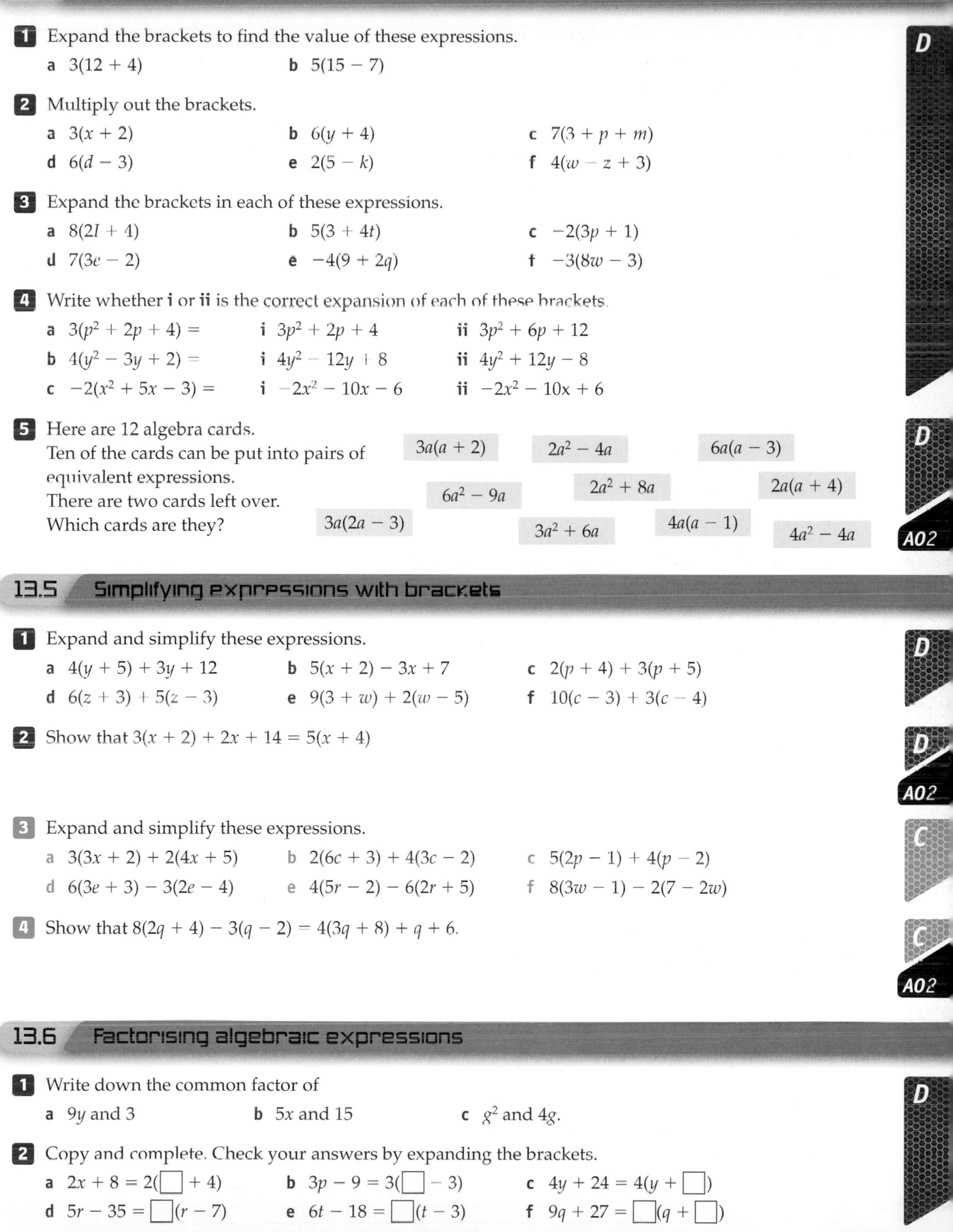

13.4 Expanding brackets

1 Expand the brackets to find the value of these expressions.

a $3(12 + 4)$ b $5(15 - 7)$

2 Multiply out the brackets.

a $3(x + 2)$ b $6(y + 4)$ c $7(3 + p + m)$

d $6(d - 3)$ e $2(5 - k)$ f $4(w - z + 3)$

3 Expand the brackets in each of these expressions.

a $8(2l + 4)$ b $5(3 + 4t)$ c $-2(3p + 1)$

d $7(3e - 2)$ e $-4(9 + 2q)$ f $-3(8w - 3)$

4 Write whether **i** or **ii** is the correct expansion of each of these brackets.

a $3(p^2 + 2p + 4) =$ **i** $3p^2 + 2p + 4$ **ii** $3p^2 + 6p + 12$

b $4(y^2 - 3y + 2) =$ **i** $4y^2 - 12y + 8$ **ii** $4y^2 + 12y - 8$

c $-2(x^2 + 5x - 3) =$ **i** $-2x^2 - 10x - 6$ **ii** $-2x^2 - 10x + 6$

5 Here are 12 algebra cards.
Ten of the cards can be put into pairs of equivalent expressions.
There are two cards left over.
Which cards are they?

$3a(a + 2)$ | $2a^2 - 4a$ | $6a(a - 3)$ | $6a^2 - 9a$ | $2a^2 + 8a$ | $2a(a + 4)$ | $3a(2a - 3)$ | $3a^2 + 6a$ | $4a(a - 1)$ | $4a^2 - 4a$

13.5 Simplifying expressions with brackets

1 Expand and simplify these expressions.

a $4(y + 5) + 3y + 12$ b $5(x + 2) - 3x + 7$ c $2(p + 4) + 3(p + 5)$

d $6(z + 3) + 5(z - 3)$ e $9(3 + w) + 2(w - 5)$ f $10(c - 3) + 3(c - 4)$

2 Show that $3(x + 2) + 2x + 14 = 5(x + 4)$

3 Expand and simplify these expressions.

a $3(3x + 2) + 2(4x + 5)$ b $2(6c + 3) + 4(3c - 2)$ c $5(2p - 1) + 4(p - 2)$

d $6(3e + 3) - 3(2e - 4)$ e $4(5r - 2) - 6(2r + 5)$ f $8(3w - 1) - 2(7 - 2w)$

4 Show that $8(2q + 4) - 3(q - 2) = 4(3q + 8) + q + 6$.

13.6 Factorising algebraic expressions

1 Write down the common factor of

a $9y$ and 3 b $5x$ and 15 c g^2 and $4g$.

2 Copy and complete. Check your answers by expanding the brackets.

a $2x + 8 = 2(\square + 4)$ b $3p - 9 = 3(\square - 3)$ c $4y + 24 = 4(y + \square)$

d $5r - 35 = \square(r - 7)$ e $6t - 18 = \square(t - 3)$ f $9q + 27 = \square(q + \square)$

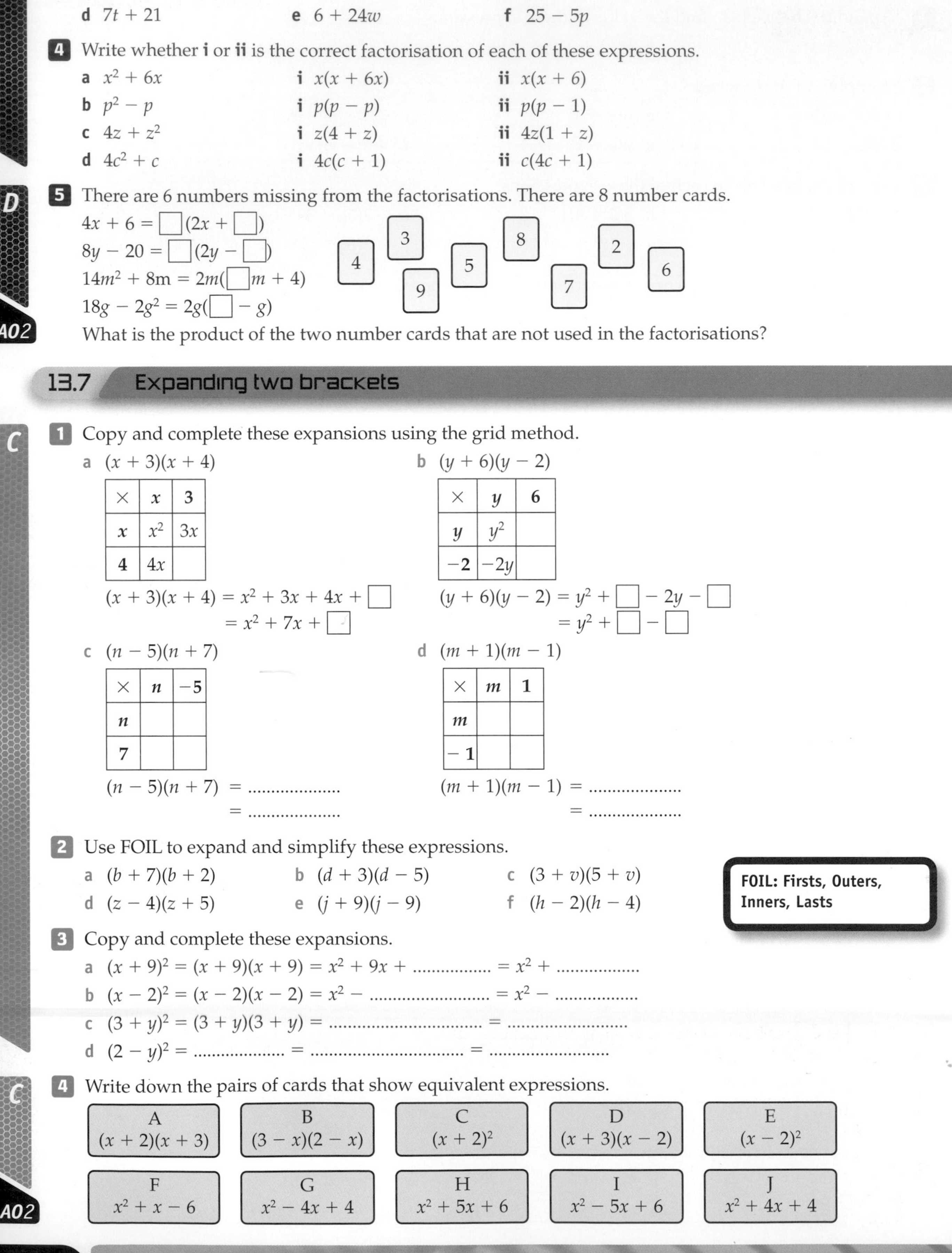

D

3 Factorise these expressions.

a $3x + 9$ b $4y - 12$ c $8z + 16$

d $7t + 21$ e $6 + 24w$ f $25 - 5p$

4 Write whether **i** or **ii** is the correct factorisation of each of these expressions.

a $x^2 + 6x$ **i** $x(x + 6x)$ **ii** $x(x + 6)$

b $p^2 - p$ **i** $p(p - p)$ **ii** $p(p - 1)$

c $4z + z^2$ **i** $z(4 + z)$ **ii** $4z(1 + z)$

d $4c^2 + c$ **i** $4c(c + 1)$ **ii** $c(4c + 1)$

D

5 There are 6 numbers missing from the factorisations. There are 8 number cards.

$4x + 6 = \square(2x + \square)$

$8y - 20 = \square(2y - \square)$

$14m^2 + 8m = 2m(\square m + 4)$

$18g - 2g^2 = 2g(\square - g)$

3 8 2 4 5 6 9 7

AO2

What is the product of the two number cards that are not used in the factorisations?

13.7 Expanding two brackets

C

1 Copy and complete these expansions using the grid method.

a $(x + 3)(x + 4)$

×	x	**3**
x	x^2	$3x$
4	$4x$	

$(x + 3)(x + 4) = x^2 + 3x + 4x + \square$
$= x^2 + 7x + \square$

b $(y + 6)(y - 2)$

×	y	**6**
y	y^2	
−2	$-2y$	

$(y + 6)(y - 2) = y^2 + \square - 2y - \square$
$= y^2 + \square - \square$

c $(n - 5)(n + 7)$

×	n	**−5**
n		
7		

$(n - 5)(n + 7) =$ ....
$=$ ....

d $(m + 1)(m - 1)$

×	m	**1**
m		
− 1		

$(m + 1)(m - 1) =$ ....
$=$ ....

2 Use FOIL to expand and simplify these expressions.

a $(b + 7)(b + 2)$ b $(d + 3)(d - 5)$ c $(3 + v)(5 + v)$

d $(z - 4)(z + 5)$ e $(j + 9)(j - 9)$ f $(h - 2)(h - 4)$

FOIL: Firsts, Outers, Inners, Lasts

3 Copy and complete these expansions.

a $(x + 9)^2 = (x + 9)(x + 9) = x^2 + 9x +$ .... $= x^2 +$ ....

b $(x - 2)^2 = (x - 2)(x - 2) = x^2 -$ .... $= x^2 -$ ....

c $(3 + y)^2 = (3 + y)(3 + y) =$ .... $=$ ....

d $(2 - y)^2 =$ .... $=$ .... $=$ ....

C

AO2

4 Write down the pairs of cards that show equivalent expressions.

A	B	C	D	E
$(x + 2)(x + 3)$	$(3 - x)(2 - x)$	$(x + 2)^2$	$(x + 3)(x - 2)$	$(x - 2)^2$

F	G	H	I	J
$x^2 + x - 6$	$x^2 - 4x + 4$	$x^2 + 5x + 6$	$x^2 - 5x + 6$	$x^2 + 4x + 4$

Links to:
Foundation Student Book
Ch14, pp. 249–264

Key Points

Finding equivalent fractions G

You make an equivalent fraction when you multiply or divide the numerator and denominator by the same number.

For example,

$\frac{1}{3} = \frac{1 \times 2}{3 \times 2} = \frac{2}{6}$

$\frac{3}{9} = \frac{3 \div 3}{9 \div 3} = \frac{1}{3}$

Writing mixed numbers and improper fractions F

In an improper fraction the numerator is larger than the denominator, for example $\frac{5}{4}$.

A mixed number has a whole number and a fractional part, for example $1\frac{1}{4}$.

Comparing two or more fractions F E D

To compare fractions with different denominators, change them into equivalent fractions with the same denominator. For example, $\frac{3}{4} = \frac{15}{20}, \frac{4}{5} = \frac{16}{20}$, so $\frac{3}{4} < \frac{4}{5}$

Multiplying fractions and whole numbers E

To multiply fractions, multiply the numerators together and multiply the denominators together.

For example, $\frac{^2\not{4}}{_1\not{5}} \times \frac{\not{15}^3}{\not{22}_{11}} = \frac{2}{1} \times \frac{3}{11} = \frac{2 \times 3}{1 \times 11} = \frac{6}{11}$

Adding and subtracting fractions G F E D

You can add or subtract fractions only when they have the same denominator. You can use equivalent fractions to do this. For example, $\frac{1}{2} + \frac{1}{3} = \frac{3}{6} + \frac{2}{6} = \frac{5}{6}$

Multiplying fractions and mixed numbers D C

To multiply mixed numbers, change them to improper fractions first.

For example, $4\frac{1}{3} \times \frac{2}{5} = \frac{13}{3} \times \frac{2}{5} = \frac{13 \times 2}{3 \times 5} = \frac{26}{15} = 1\frac{11}{15}$

Dividing by a fraction D C

To divide by a fraction, turn the fraction upside down and multiply. When the division involves mixed numbers, change them to improper fractions first.

For example,

$\frac{3}{4} \div 7 = \frac{3}{4} \div \frac{7}{1} = \frac{3}{4} \times \frac{1}{7} = \frac{3 \times 1}{4 \times 7} = \frac{3}{28}$

Adding and subtracting mixed numbers C

You can add or subtract mixed numbers by changing them into improper fractions.

For example,

$1\frac{1}{3} + 3\frac{2}{5} = \frac{4}{3} + \frac{17}{5} = \frac{20}{15} + \frac{51}{15} = \frac{71}{15} = 4\frac{11}{15}$

Finding reciprocals C

When two numbers can be multiplied together to give an answer of 1, then each number is called the reciprocal of the other.

The reciprocal of a fraction is found by turning the fraction upside down.

The reciprocal of a number is 1 divided by that number

14.1 Comparing fractions

G

1 Write down five fractions that are equivalent to $\frac{1}{4}$.

2 Write down five fractions that are equivalent to $\frac{2}{3}$.

F

3 Write $\frac{1}{4}$ and $\frac{3}{10}$ as equivalent fractions with a denominator of 20.
Which fraction is larger?

4 Write $\frac{1}{3}$, $\frac{4}{9}$ and $\frac{1}{2}$ as equivalent fractions with a denominator of 18.
Which fraction is the smallest?

E AO2

5 Gupta says that $\frac{5}{12}$ is smaller than $\frac{7}{18}$.
Is Gupta correct?
Show how you decided.

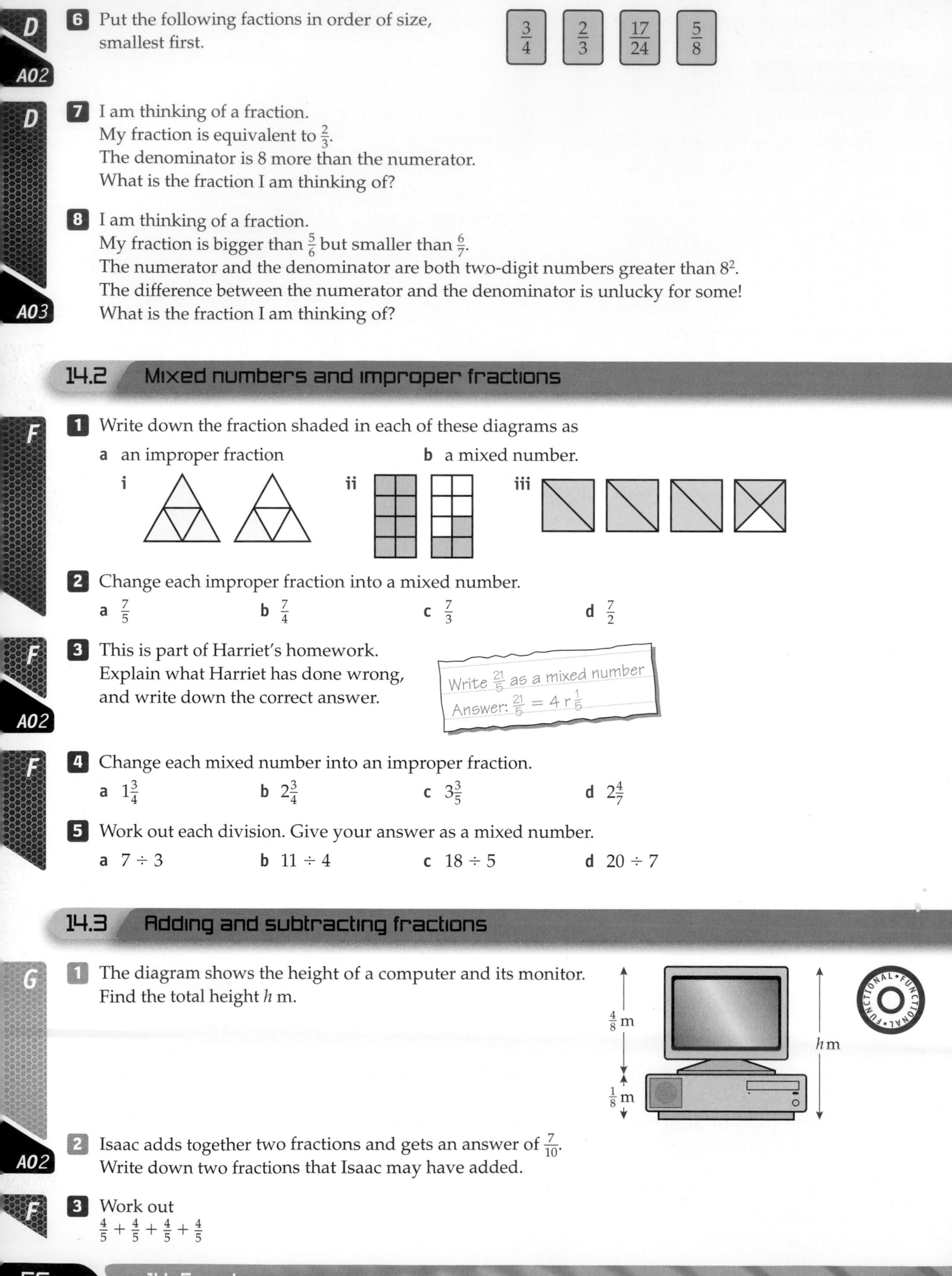

D A02

6 Put the following factions in order of size, smallest first.

$\frac{3}{4}$ $\frac{2}{3}$ $\frac{17}{24}$ $\frac{5}{8}$

D A03

7 I am thinking of a fraction.
My fraction is equivalent to $\frac{2}{3}$.
The denominator is 8 more than the numerator.
What is the fraction I am thinking of?

8 I am thinking of a fraction.
My fraction is bigger than $\frac{5}{6}$ but smaller than $\frac{6}{7}$.
The numerator and the denominator are both two-digit numbers greater than 8^2.
The difference between the numerator and the denominator is unlucky for some!
What is the fraction I am thinking of?

14.2 Mixed numbers and improper fractions

F

1 Write down the fraction shaded in each of these diagrams as

a an improper fraction **b** a mixed number.

i **ii** **iii**

2 Change each improper fraction into a mixed number.

a $\frac{7}{5}$ **b** $\frac{7}{4}$ **c** $\frac{7}{3}$ **d** $\frac{7}{2}$

F A02

3 This is part of Harriet's homework.
Explain what Harriet has done wrong, and write down the correct answer.

F

4 Change each mixed number into an improper fraction.

a $1\frac{3}{4}$ **b** $2\frac{3}{4}$ **c** $3\frac{3}{5}$ **d** $2\frac{4}{7}$

5 Work out each division. Give your answer as a mixed number.

a $7 \div 3$ **b** $11 \div 4$ **c** $18 \div 5$ **d** $20 \div 7$

14.3 Adding and subtracting fractions

G

1 The diagram shows the height of a computer and its monitor.
Find the total height h m.

A02

2 Isaac adds together two fractions and gets an answer of $\frac{7}{10}$.
Write down two fractions that Isaac may have added.

F

3 Work out
$\frac{4}{5} + \frac{4}{5} + \frac{4}{5} + \frac{4}{5}$

4 Jess adds together two identical fractions and gets an answer of $1\frac{13}{15}$.
Write down the fractions that Jess added.

5 Work out

a $\frac{8}{9} - \frac{1}{3}$ b $\frac{1}{18} + \frac{1}{6}$

c $\frac{11}{12} - \frac{3}{4}$ d $\frac{21}{100} + \frac{3}{10}$

6 Kim is tying bags of almonds for her friend's wedding.
Each bag uses $\frac{3}{20}$ m of gold ribbon.
Kim needs enough ribbon for 80 bags of almonds.
The ribbon costs £2.50 per metre.
What is the total cost of the ribbon?

7 Work out

a $\frac{1}{4} + \frac{1}{3}$ b $\frac{1}{4} - \frac{1}{5}$

c $\frac{4}{5} + \frac{1}{3}$ d $\frac{3}{5} - \frac{1}{4}$

14.4 Adding and subtracting mixed numbers

1 Work out

a $2\frac{3}{5} + 1\frac{1}{4}$ b $5\frac{1}{2} - 1\frac{3}{5}$

c $1\frac{2}{3} + 3\frac{1}{2} + 1\frac{3}{4}$ d $1\frac{1}{2} + 1\frac{1}{5} + 1\frac{1}{6}$

2 Kayla buys a $\frac{1}{2}$ kg bag of porridge, a $1\frac{1}{8}$ kg salmon, a $2\frac{3}{4}$ kg turkey and some fruit weighing a total of $1\frac{1}{2}$ kg.
She puts all her shopping into a box, which weighs $\frac{1}{4}$ kg.
What is the total weight of her shopping and the box?

3 Mark is fitting a new worktop in his kitchen.
He buys a worktop which is $3\frac{1}{5}$ m long.
He cuts off the amount he needs and has $1\frac{1}{8}$ m left over.
What length of worktop did Mark need?

4 Nigel is making a model tower from balsa wood.
The instructions for the model are given in inches.
Nigel has to make four different parts and glue them together.
The base is $3\frac{1}{8}$ inches high.
The two middle parts are $1\frac{3}{4}$ inches and $\frac{11}{16}$ inches high.
The top part is $2\frac{1}{2}$ inches high.
What is the total height of the tower?

5 A ferry has a maximum weight limit of 44 tonnes for any vehicle it transports.
The front end of a lorry weighs $5\frac{1}{4}$ tonnes and its trailer weighs $2\frac{4}{5}$ tonnes.
The lorry is carrying a container.
The container weighs $2\frac{9}{10}$ tonnes and its contents weighs $29\frac{7}{10}$ tonnes.
Show that the lorry will be allowed on the ferry.

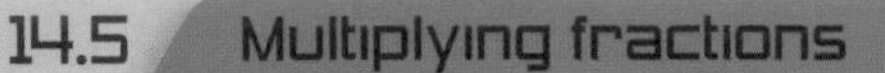

14.5 Multiplying fractions

1 Which is the larger amount, $\frac{4}{5}$ of £180 or $\frac{1}{7}$ of £1015?

2 The fraction of a ripe cucumber that is water is $\frac{19}{20}$.
How much of a 280 g cucumber is **not** water?

3 Olive is driving from the UK to Spain.
The trip is 1240 miles.
So far she estimates she has driven $\frac{3}{8}$ of the way.
How many miles further does Olive have to go?

AO2

4 Work out

a $\frac{1}{4} \times \frac{3}{5}$ b $\frac{3}{5} \times \frac{4}{7}$ c $\frac{2}{5} \times \frac{10}{12}$ d $\frac{3}{8} \times \frac{8}{9}$

5 Priya buys a length of wood and uses $\frac{5}{8}$ of it for a shelf.
She then uses $\frac{2}{5}$ of what is left to make a smaller shelf.
The rest of the wood she burns on her fire.
What fraction of the length of wood does Priya burn?

AO3

14.6 Multiplying mixed numbers

1 Which is the larger number, $7 \times 1\frac{3}{4}$ or $2\frac{1}{8} \times 6$?

2 Quinn packs 7 identical laptops in a box.
Each laptop weighs $5\frac{3}{5}$ kg. The box weighs $2\frac{1}{2}$ kg.
What is the total weight of the laptops and box?

AO2

3 Work out these multiplications.

a $\frac{3}{7} \times 4\frac{5}{6}$ b $5\frac{1}{4} \times \frac{8}{11}$ c $1\frac{11}{12} \times \frac{10}{11}$

4 Rhys is going to fit some laminate flooring.
Each strip of flooring is $\frac{3}{8}$ m wide.
The length of flooring Rhys needs is $43\frac{1}{2}$ m.

a What area of laminate flooring does Rhys need?

The laminate flooring costs £37.50 per square metre.
It can only be bought in a whole number of square metres.

b How much does Rhys pay for the laminate flooring?

Area of rectangle = length × width

5 Sam is a home worker. She packs information leaflets into envelopes.
Sam packs, on average, a dozen envelopes in $3\frac{1}{2}$ minutes.
Sam must pack 1080 envelopes in a day to get a full day's pay.
Approximately how long will it take Sam to earn a full day's pay?

First find $\frac{1}{12}$ of $3\frac{1}{2}$ minutes to find out how long it takes Sammy to pack one envelope.

AO3

14.7 Reciprocals

1 a Write down the reciprocal of 7.
b What do you get if you multiply 7 by its reciprocal?

2 a Write down the reciprocal of 200.
b What do you get if you multiply 200 by its reciprocal?

3 a Write down the reciprocal of $\frac{1}{9}$.
b What do you get if you multiply $\frac{1}{9}$ by its reciprocal?

4 a Write down the reciprocal of $\frac{2}{9}$.
b What do you get if you multiply $\frac{2}{9}$ by its reciprocal?

5 a Work out the reciprocal of 0.3.
b What do you get if you multiply 0.3 by its reciprocal?

6 a Work out the reciprocal of 0.05.
b What do you get if you multiply 0.05 by its reciprocal?

7 a Work out the reciprocal of 2.5.
b What do you get if you multiply 2.5 by its reciprocal?

14.8 Dividing fractions

1 Work out

a $\frac{3}{5} \div 8$ **b** $\frac{3}{5} \div 9$ **c** $8 \div \frac{3}{5}$ **d** $9 \div \frac{3}{5}$

2 What do you notice about your answers to **1a** and **1c**?

3 Work out

a $\frac{3}{5} \div \frac{5}{8}$ **b** $\frac{6}{11} \div \frac{12}{13}$ **c** $3\frac{2}{7} \div 10$ **d** $12\frac{3}{10} \div 9$

4 Tyrone pours $\frac{7}{8}$ of a litre of engine oil into bottles that hold $\frac{1}{10}$ of a litre.
How many bottles can he fill?

5 Ulrik has a blank music CD that will play music for 57 minutes.
He burns his favourite song onto the CD as many times as he can.
The song lasts $3\frac{4}{5}$ minutes.
How many copies of the song can he fit onto the CD?

6 Vicky shares $2\frac{1}{2}$ chocolate cakes among 8 people.
What fraction of a cake do they each receive?

Links to:
Foundation Student Book
Ch15, pp. 265–277

Key Points

Place value headings for numbers F G

The decimal point separates the whole number part from the fraction part of a number.

Whole numbers				Fractions		
Hundreds	Tens	Units	.	tenths	hundredths	thousandths
100	10	1	.	$\frac{1}{10}$	$\frac{1}{100}$	$\frac{1}{1000}$

Multiplying and dividing decimals by a power of 10 F

To multiply:

- by 10, move each digit one place to the left
- by 100, move the digits two places to the left
- by 1000, move the digits three places to the left.

To divide:

- by 10, move the digits one place to the right
- by 100, move the digits two places to the right
- by 1000, move the digits three places to the right.

Adding and subtracting decimals E

To add and subtract decimals, set out the calculation in columns by lining up the decimal points, then add or subtract.

Converting decimals to fractions D

You can use the place value headings to convert a decimal to a fraction.

First write the decimal as tenths, hundredths, thousandths, ...

Take the biggest denominator and put it on the bottom of a fraction. Put the decimal digits on the top of the fraction. Remember to give your answer in its simplest form.

Multiplying decimals D C

To multiply decimals:

- ignore the decimal points and just multiply the numbers
- count the decimal places in the calculation
- put this number of decimal places in the answer.

Dividing decimals D C

You can divide a decimal by a whole number in the usual way.

If you are dividing by a decimal, first write it as a fraction, and convert the denominator to a whole number. Then divide in the usual way.

Terminating decimals D

Terminating decimals are decimals that come to an end. For example, 0.6.

Recurring decimals C

Recurring decimals are decimals that never come to an end, for example 0.333...

15.1 Multiplying and dividing by 10, 100, 1000 ...

F

1 Work out

a 123×10 **b** 12.3×10 **c** 1.23×10 **d** 0.123×10

e $123 \div 10$ **f** $12.3 \div 10$ **g** $1.23 \div 10$ **h** $0.123 \div 10$

2 Work out

a 1230×100 **b** 1.23×1000

c $12.34 \div 100$ **d** $123.4 \div 1000$

e $700 \div 100$ **f** $1230 \div 10000$

g 1000×123.4 **h** $0.0123 \div 100$

3 **a** What has 0.5 been multiplied by to give 500?

b What has 50 been multiplied by to give 5000?

c What has 5.5 been divided by to give 0.055?

4 Choose a number and put it through this number machine.
What do you notice about your answer?
Explain why this happens.

→ ×10 → ÷10 → ×100 → ÷100 →

F

5 A factory in Brazil produces bottles of cola containing 0.33 litre.
They produce 10 000 bottles per day.
How many litres of cola do they produce per day?

FUNCTIONAL

AO2

15.2 Adding and subtracting decimals

E

1 Walt has emptied his piggy-bank.
He has £0.93 in 1p coins, £1.22 in 2ps, £1.65 in 5ps, £3.80 in 10ps and £0.80 in 20p.

a How much money does Walt have altogether?

b How much more money does Walt need to buy a shirt costing £10?

FUNCTIONAL

2 Xiao Chen weighs 96.8 kg.
He needs to increase his weight to 102.5 kg to be able to enter a body-building competition.
How much weight does Xiao Chen need to put on?

FUNCTIONAL

3 Work out 6.31 + 12.8 + 130.

E

4 The diagram shows the distances between markers on an orienteering course.
Mel runs in a straight line from the start at (1) to (2) to (3) to (4) and back to (1).
What is the total length that Mel runs?

(1) to (2): 0.96 km
(2) to (3): 1.71 km
(3) to (4): 1.88 km
(4) to (1): 1.32 km

FUNCTIONAL

5 A plank of wood is 85 cm long.
Three identical pieces are cut from the plank.
Each piece is 22.6 cm long.
How long is the piece of wood that is left?

FUNCTIONAL

6 Here are four numbers.
Which two numbers are the closest together?

10 14.8 19.76 24.44

AO2

15.3 Converting decimals to fractions

D

1 Convert each of these decimals to fractions in their lowest terms.

a 0.2	b 0.4	c 0.5	d 0.7
e 0.25	f 0.28	g 0.08	h 0.85
i 0.125	j 0.875	k 0.055	l 0.004

2 Convert each of these decimals to mixed numbers.

a 2.2	b 3.4	c 4.5	d 8.7
e 6.28	f 9.05	g 12.666	h 2.0001

3 Dylan carried out a survey and found that the proportion of students in his year group that had exactly one brother or sister was 0.52.
What fraction of his year group had exactly one brother or sister?
Write your answer in its lowest terms.

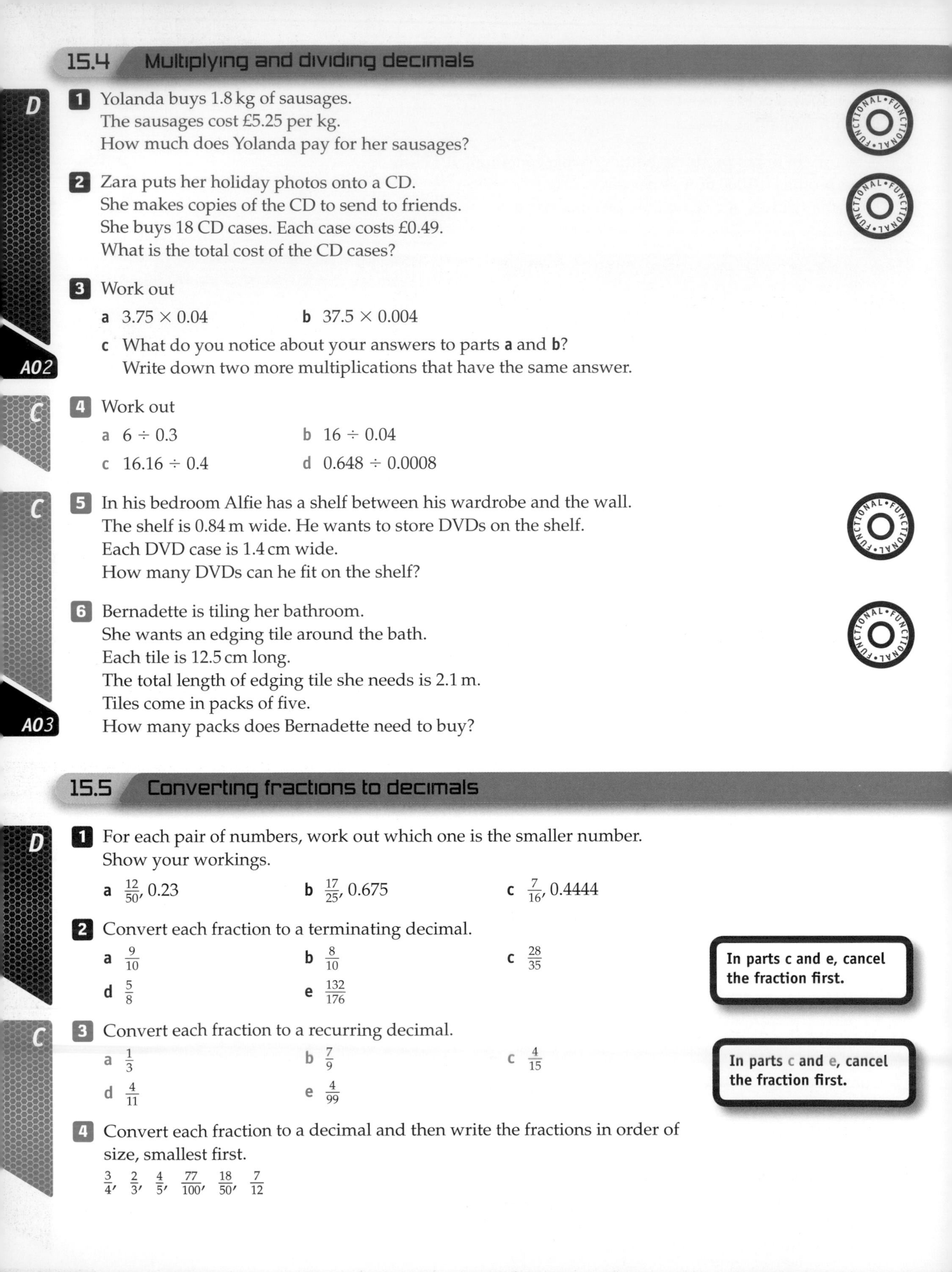

15.4 Multiplying and dividing decimals

D

1 Yolanda buys 1.8 kg of sausages.
The sausages cost £5.25 per kg.
How much does Yolanda pay for her sausages?

FUNCTIONAL

2 Zara puts her holiday photos onto a CD.
She makes copies of the CD to send to friends.
She buys 18 CD cases. Each case costs £0.49.
What is the total cost of the CD cases?

FUNCTIONAL

3 Work out

a 3.75×0.04 **b** 37.5×0.004

c What do you notice about your answers to parts **a** and **b**?
Write down two more multiplications that have the same answer.

AO2

C

4 Work out

a $6 \div 0.3$ b $16 \div 0.04$

c $16.16 \div 0.4$ d $0.648 \div 0.0008$

C

5 In his bedroom Alfie has a shelf between his wardrobe and the wall.
The shelf is 0.84 m wide. He wants to store DVDs on the shelf.
Each DVD case is 1.4 cm wide.
How many DVDs can he fit on the shelf?

FUNCTIONAL

6 Bernadette is tiling her bathroom.
She wants an edging tile around the bath.
Each tile is 12.5 cm long.
The total length of edging tile she needs is 2.1 m.
Tiles come in packs of five.
How many packs does Bernadette need to buy?

FUNCTIONAL

AO3

15.5 Converting fractions to decimals

D

1 For each pair of numbers, work out which one is the smaller number.
Show your workings.

a $\frac{12}{50}$, 0.23 **b** $\frac{17}{25}$, 0.675 **c** $\frac{7}{16}$, 0.4444

2 Convert each fraction to a terminating decimal.

a $\frac{9}{10}$ **b** $\frac{8}{10}$ **c** $\frac{28}{35}$

d $\frac{5}{8}$ **e** $\frac{132}{176}$

In parts c and e, cancel the fraction first.

C

3 Convert each fraction to a recurring decimal.

a $\frac{1}{3}$ b $\frac{7}{9}$ c $\frac{4}{15}$

d $\frac{4}{11}$ e $\frac{4}{99}$

In parts c and e, cancel the fraction first.

4 Convert each fraction to a decimal and then write the fractions in order of size, smallest first.

$\frac{3}{4}$, $\frac{2}{3}$, $\frac{4}{5}$, $\frac{77}{100}$, $\frac{18}{50}$, $\frac{7}{12}$

Links to:
Foundation Student Book
Ch16, pp. 278–292

Key Points

Equations as balanced scales F

Use inverse operations to solve equations.

For example, to solve $x + 3 = 10$, subtract 3 from both sides of the equation:

$x + 3 - 3 = 10 - 3$

$x = 7$

Solving-two step equations E D

You will need to use two or more steps to solve some equations.

For example, to solve $2x + 5 = 11$:

$2x + 5 - 5 = 11 - 5$

$2x = 6$

$2x \div 2 = 6 \div 2$

$x = 3$

Writing and solving equations E D C

When writing equations, decide what your unknown will represent and then write an expression. Turn this into an equation by looking for expressions or numbers that are equal.

Equations with brackets D

When you solve equations with brackets, expand the brackets first.

Equations with an unknown on both sides D C

To solve this type of equation, you need to get the unknown on one side of the equals sign only.

For example, $3x - 3 = 2x + 5$.

$3x - 2x - 3 = 2x - 2x + 5$

$x - 3 = 5$

Then solve the equation as normal.

Inequalities and the number line E D C

$x > 2$ is an inequality. It means that x must be greater than 2. It can be shown on a number line.

0 1 2 3 4 5 6 7

The open circle means that the 2 is not included.

$x \leq 5$ means that x is less than or equal to 5. The closed circle shows that in this case the 5 is included in the set of solutions.

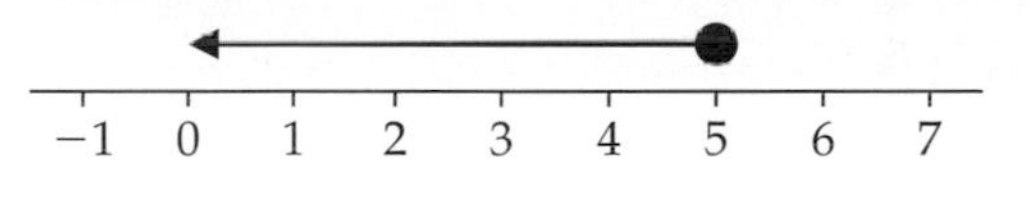

Using the balance method to solve inequalities C

You can use the balance method to solve inequalities.

16.1 Solving simple equations

1 Solve the following equations:

- **a** $a + 5 = 15$
- **b** $b + 22 = 26$
- **c** $c + 100 = 320$
- **d** $d - 8 = 12$
- **e** $e - 20 = 70$
- **f** $f - 35 = 35$
- **g** $5 + g - 15$
- **h** $22 + h = 26$
- **i** $20 + i = 320$

2 Solve the following equations:

- **a** $j - 5 = -2$
- **b** $k - 12 = -2$
- **c** $m + 50 = -2$
- **d** $2n - 10$
- **e** $9p = 81$
- **f** $30 = 5q$

3 Solve the following equations:

- **a** $10r = -30$
- **b** $6s = -18$
- **c** $\frac{t}{2} = -6$

16.2 Solving two-step equations

E

1 Draw a set of scales and work out the value of x in this equation.
$5x + 2 = 17$

2 Solve these equations.

a $2a + 3 = 23$ b $3b + 2 = 14$ c $4c - 5 = 15$
d $2d + 3 = -5$ e $3e - 2 = -20$ f $18 - 2f = 12$

D

3 Solve the following:

a $\frac{a}{2} + 6 = 10$ b $\frac{b+6}{2} = 10$ c $\frac{c}{4} - 5 = 5$
d $\frac{d-5}{4} = 5$ e $7 + \frac{e}{3} = 10$ f $\frac{7+f}{3} = 10$

4 Solve the following:

a $5a - 12 = -2$ b $\frac{b}{2} - 10 = -8$ c $\frac{c-3}{5} = 2$

5 Solve the following:

a $5a = 7$ b $3b - 2 = 6$ c $4c + 2 = -7$
d $6d - 2 = 1$ e $5 = 3e + 4$ f $8 = 8f + 3$

16.3 Writing and solving equations

E

1 Amin thinks of a number and divides it by 3. His answer is 5.
Use algebra to work out Amin's original number.

AO3

2 Oscar thinks of a number and multiplies it by 2. His answer is 100.
Use algebra to work out Oscar's original number.

D

3 Josh is three times the age of his brother Tim.

a Use the letter x to represent Tim's age.
Write an expression for Josh's age in terms of x.

b Write an expression for the total age of Josh and Tim.

c The total of their ages is 12.
Write an equation and solve it to find out how old Tim is.

4 a Write an expression for the perimeter of this trapezium.

b Simplify your expression in part **a**.

c The perimeter of the trapezium is 26 cm.
Write an equation and solve it to find the value of x.

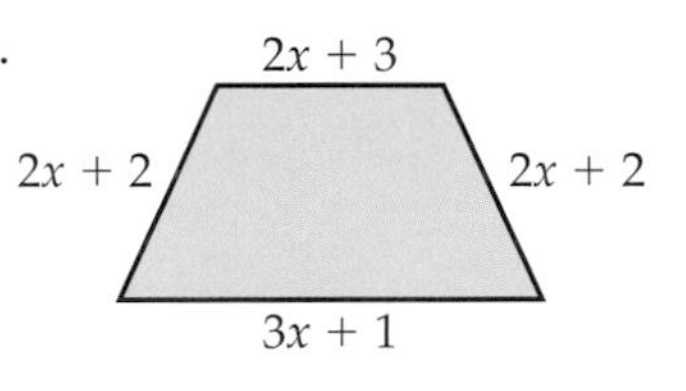

The perimeter is the distance around the outside of a shape.

5 a Explain why the sum of four consecutive numbers can be written as
$x + x + 1 + x + 2 + x + 3$.

b Simplify this expression.

c The sum of four consecutive numbers is 74.
Use part **b** to write an equation, and solve it to find the value of the first number.

AO2

C

6 Pauline is going on holiday. The total cost of the holiday is £488.
She pays a deposit of 25 per cent of the total cost.
In 12 weeks Pauline must pay the remainder of the money.
Let x be the amount of money that she must save each week for the holiday.
Write an equation and solve it to find the amount of money Pauline must save each week.

AO2

16.4 Equations with brackets

1 Solve the following:

a $3(a + 4) = 15$ **b** $2(b + 5) = 22$ **c** $4(c + 4) = 20$

d $2(d - 5) = 4$ **e** $3(2e - 5) = 3$ **f** $4(2f + 5) = 52$

g $2(2g - 3) = 2$ **h** $24 = 3(3h + 2)$ **i** $-5 = 5(3i - 4)$

2 Solve the following:

a $3(a + 2) - 2 = 19$ b $2(3b - 2) + 7 = 27$ c $5(c + 8) - 25 = 25$

d $2(d + 8) + 3d = 36$ e $3(4e - 5) - 3e = 21$ f $4(3f - 6) + 8 - 8f = 4$

16.5 Equations with an unknown on both sides

1 Solve the following:

a $6x + 5 = 3x + 14$ **b** $9x + 1 = 5x + 9$ **c** $9x - 11 = 3x + 13$

d $8x - 12 = 5x - 3$ **e** $16x - 35 = 7x - 8$ **f** $11x - 51 = 6x - 1$

g $4x + 7 = 6x - 5$ **h** $7x - 1 = 12x - 16$ **i** $2x + 15 = 5x + 3$

j $4x + 3 = 21 - 2x$ **k** $x - 5 = 7 - x$ **l** $3x - 16 = -1 - 2x$

2 Solve the following:

a $8x + 8 = 2(x + 16)$ b $5(x + 4) = 3(x + 12)$ c $3(x - 4) = 8(x - 9)$

16.6 Inequalities and the number line

1 Rewrite this maths sentence in words.
$33 > 27$

2 Write down whether each inequality is true or false.

a $27 < 33$ b $7 > 6$ c $8 \leqslant 10$ d $12 \geqslant 12.5$

3 Show each inequality on a number line.

a $x > 7$ b $x < 7$ c $x \geqslant 7$

d $x \leqslant 5$ e $-2 < x \leqslant 3$ f $x \leqslant 0$ or $x > 3$

Check that you have filled in the circle for Q3c.

4 Write down all the whole number values for x in each inequality.

a $0 < x \leqslant 5$ **b** $-3 \leqslant x < 2$ **c** $-7 < x < -1$

16.7 Solving inequalities

1 Solve each of the following inequalities.

a $5x > 30$ b $\frac{x}{10} \geqslant 30$ c $x \quad 7 \leqslant 30$

2 Solve each of the following inequalities.

a $5x - 15 > 30$ b $4x + 2 \geqslant 30$ c $\frac{x}{3} + 23 \leqslant 30$

d $17x > 16x - 30$ e $30 \leqslant 6x + 6$ f $3x > 5x + 8$

3 Solve each inequality and show the results on a number line.

a $2x \leqslant -8$ b $5x > 2x - 6$ c $-15 < 5x \leqslant 5$

Links to: Foundation Student Book Ch17, pp. 293–310

Key Points

Formulae given in words G

A formula is a general rule that shows the relationship between quantities that can vary.

For example,
pay = hours worked × rate of pay.

Formulae written using letters and symbols F E D

You can use letters for the variables in a formula.

For example, $p = hr + b$
where p is the pay, h is the hours worked, r is the rate of pay and b is the bonus.

Substitution

Use the correct order of operations to help you do the calculations when you substitute values into an algebraic expression or a formula.

Indices E D C

x^2 is called 'x squared', y^3 is called 'y cubed'.
$t \times t \times t \times t = t^4$. You say '$t$ to the power 4'.

The ^{4} is called the index or power.

Law of indices C

Any number or variable raised to the power 1 is equal to the number or variable itself.

For example: $3^1 = 3$, $x^1 = x$

To *multiply* powers of the *same* number or variable, *add* the indices.

For example: $4^2 \times 4^3 = 4^{2+3} = 4^5$, $x^n \times x^m = x^{n+m}$

To *divide* powers of the *same* number or variable, *subtract* the indices.

For example: $5^4 \div 5^2 = 5^{4-2} = 5^2$, $x^n \div x^m = x^{n-m}$

Changing the subject of a formula C

The subject of a formula only appears once, and only on its own side of the formula.

In the formula $v = u + at$, the variable v is the subject.

You can rearrange a formula to make a different variable the subject.

You can rearrange $v = u + at$ as $a = \frac{(v - u)}{t}$

17.1 Using formulae

G

1 Andrew works out the total cost of the electricity he uses, using this formula:

total cost = number of units used × cost per unit.

Work out the total cost of the electricity Andrew uses when

a he uses 1500 units and the cost per unit is £0.08

b he uses 2000 units and the cost per unit is £0.07.

2 Mairead works out her total electricity bill, using this formula:

electricity bill = number of units used × cost per unit + standing charge.

Work out Mairead's total electricity bill when

a she uses 1200 units, the cost per unit is £0.07 and the standing charge is £37.50

b she uses 2150 units, the cost per unit is £0.095 and the standing charge is £15.25.

F

3 The formula for the area of a parallelogram is:

A = bh, where b is the base length and h is the height.

Work out the area of a parallelogram with a base length of 9 cm and a height of 5 cm.

4 A formula linking force (F), mass (M) and acceleration (a) is:

$F = Ma$

Work out the value of F when $M = 6$ and $a = 4$.

5 A formula to work out the length of time, T minutes, to cook a joint of meat is:

$T = M \times 80 + 30$, where M is the mass of the joint of meat in kg.

Work out the time it will take to cook a joint of meat weighing 2 kg.
Give your answer in hours and minutes.

6 Cath uses this formula to work out the cost of some pencils.

Cost = number of pencils × cost of one pencil

Cath spends £3.50 altogether on pencils that cost 25p each.
How many pencils does Cath buy?

FUNCTIONAL F A02

17.2 Writing your own formulae

E

1 Calculators cost £5 each.
Write a formula for the total cost, C pounds, of x calculators.

2 Blank DVDs cost b pence each.
Write a formula for the total cost, T pence, of 20 DVDs.

D

3 A chocolate egg costs a pence and a chocolate chicken costs b pence.
Savita buys 13 chocolate eggs and 3 chocolate chickens.
Write a formula for the total cost, C pence, of these items.

4 When Barry goes for a long walk, he estimates it will take him 20 minutes per mile plus an extra half an hour.
Write a formula for the total time taken, T minutes, for Barry to walk x miles.

5 When Barry goes for a very long walk, he estimates it will take him 17 minutes per mile plus an extra hour.
Write a formula for the total time taken, T minutes, for Barry to walk x miles.

6 A square has a side length of $2x + 2$ cm.
Write a formula for the perimeter, p cm, of this square.

17.3 Using index notation

E

1 Simplify the following expressions using index notation.

a $a \times a \times a$ b $b \times b \times b \times b \times b$ c $c \times c$

Remember you are multiplying not adding. $a \times a \times a$ is not the same as $3a$.

D

2 Simplify the following expressions using index notation.

a $3 \times a \times a \times a \times a$ b $b \times b \times 10 \times b \times b$ c $1 \times c \times 2 \times c \times 3 \times c$

3 Simplify the following expressions using index notation.

a $3a \times 5a$ b $2b \times 3b \times b$ c $a \times 2a \times 3a \times 2a \times a$

C

4 Simplify the following:

a $x^2 \times x^2$ b $b^3 \times b^3$ c $x^3 \times x^4$ d $d^{56} \times d^2$.

Remember you are multiplying not adding. $x^2 \times x^2$ is not the same as $2x^2$.

5 Simplify the following:

a $3a^2 \times 2a^3$ b $5x^5 \times 4x^4$ c $8c^3 \times c^4$.

C

6 Simplify the following:

a $a^4 \div a^2$ b $x^7 \div x^3$ c $c^{99} \div c^{96}$.

7 Simplify the following:

a $7x^5 \div x^3$ b $10b^4 \div 5b^2$ c $8c^3 \div 2c$

C

8 Explain how you worked out your answer to Q4d.

AO3 **9** Explain how you worked out your answer to Q6c.

17.4 Laws of indices (index laws)

C

1 Simplify each of the following expressions.

a $w^2 \times w^3$ b $x^7 \times x^4$ c $y^9 \times y$

d $2p^2 \times 5p^4$ e $4t \times 4t^3 \times t^2$ f $2r^2 \times 3r^3 \times 2r$

2 Simplify each of the following expressions.

a $a^7 \div a^4$ b $b^3 \div b$ c $c^3 \div c^3$

d $8d^2 \div 4d$ e $12e^6 \div 4e^3$ f $18f^2 \div 9f^2$

3 Simplify each of the following expressions.

a $\dfrac{w^5 \times w^3}{w^4}$ b $\dfrac{d^3 \times 4d^2}{d}$ c $\dfrac{c^5 \times 8c^3}{c^6}$

d $\dfrac{4x^3 \times 2x^3}{8x^2}$ e $\dfrac{4c^6 \times 3c^3}{3c \times 2c^3}$

17.5 Substituting into expressions

F

1 When $w = 2$, $x = 3$ and $y = 5$, work out the value of these expressions.

a $2x$ b $4y$ c $w + x + y$ d $10 - x$

e $y - w$ f wx g wxy h $xy - w$

Remember that wxy means $w \times x \times y$.

2 When $a = 3$, $b = \frac{1}{2}$ and $c = -5$, work out the value of these expressions.

a $4b$ b $3c$ c $2a + 6$ d $4a + c$

E

3 When $p = 2$, $q = 2.5$ and $r = -2$, work out the value of these expressions.

a $6p + 2q$ b $5p + 2r$ c $2p + 2q + 2r$

d $3q + 3r$ e $4p + 5q$ f $8q - 8p + 2r$

Remember that $+ -4$ is the same as -4.

D

4 When $e = 4$, $f = 0.5$ and $g = -2$, work out the value of these expressions.

a $g^2 + e$ b $2e^2 + f$ c $e^3 + g$

d $2g^2 + ef$ e $10e^2 - 100f$ f $g^3 - e$

g $4(e + 6)$ h $e(10f - 3)$ i $\dfrac{e^2 + g}{7}$

j $\dfrac{g^2 + 8f}{e}$ k $\dfrac{g^3 + g^2}{e}$ l $\dfrac{f(e + g)}{e}$

Remember that e^3 means $e \times e \times e$.

Order of operations: brackets first.

17.6 Substituting into formulae

E

1 Use the formula $I = mu - mv$ to work out the value of I when $m = 5$, $u = 8$ and $v = 3$.

2 The formula to work out the shaded area, A, of this shape is $A = lw - \frac{1}{2}bh$.
Find the shaded area when $l = 12$, $w = 10$, $b = 8$ and $h = 6$.

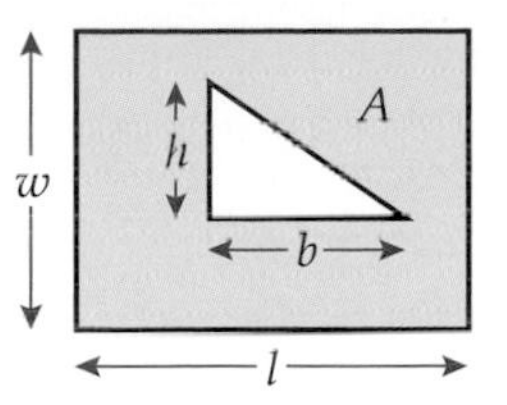

3 The formula for working out the perimeter of a regular pentagon is $P = 5l$, where l is the length of each side.
Work out the value of l when $P = 20$ cm.

E FUNCTIONAL

4 An electricity company calculates electricity bills each quarter (three months) using this formula:

$C = 0.08U + 25$ where C is the amount to pay in pounds and U is the number of units of electricity used.

A customer uses 1200 units in January, 1100 units in February and 500 units in March.

a Work out the customer's bill for the first quarter of the year.

b The customer would like her bill to be less than £200 per quarter.
How many units of electricity can she use, to have a bill of £200?

AO2

D FUNCTIONAL

5 A person's body mass index (BMI) is worked out using this formula:

BMI $= \frac{m}{h^2}$, where m is the mass in kg and h is the height in m.

Work out the BMI of a person who weighs 72 kg and is 1.83 m tall.

6 The formula for the area, A, of a parallelogram is

$A = bh$,

where b is the base measurement, and h is the perpendicular height.

Use the formula to work out

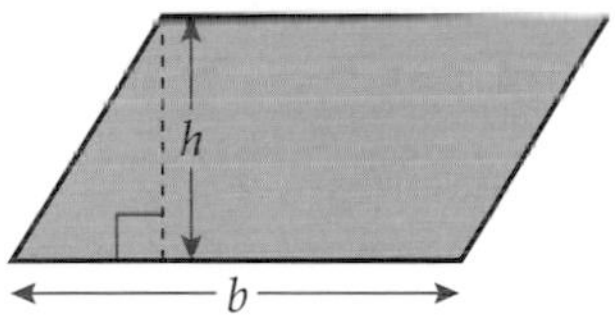

a A when $b = 6$ and $h = 4$

b b when $A = 15$ and $h = 3$

c h when $A = 42$ and $b = 7$.

17.7 Changing the subject of a formula

C

1 Make x the subject of these formulae by rearranging.

a $a = x + t$ b $p = x - c$ c $t = 3x$

d $v = ax$ e $y = 4x + 2$ f $y = 3x - 1$

g $y = mx + c$ h $T = 3x + 2y$ i $w = 6x - 4t$

j $t^2 = ax + 6$ k $v = u - 3x$ l $v = u - xy$

m $W = \frac{1}{2}x + 4$ n $v = 3(x + 2)$ p $e = t(x - y)$

18 Percentages

Links to:
Foundation Student Book
Ch18, pp. 311–326

Key Points

Credit — E

When you buy on credit, you usually have to pay a deposit followed by a number of regular payments.

Calculations involving simple interest — E

When you put money into a savings account in a bank or building society, they pay you interest. If the interest is the same amount each year, it is called simple interest.

VAT — D

Value added tax (VAT) is added to the price of items and services. Generally it is 17.5% in the UK.

Percentage increase and decrease — D

Method A

1 Work out the value of the increase (or decrease).
2 Add to (or subtract from) the original amount.

Method B

1 Add the percentage increase to 100% (or subtract the percentage decrease from 100%).
2 Convert this percentage to a decimal.
3 Multiply it by the original amount.

Percentage profit or loss — C

Percentage profit (or loss)

$$= \frac{\text{actual profit (or loss)}}{\text{cost price}} \times 100\%$$

where the actual profit (or loss) is the difference between the cost price and the selling price.

Repeated percentage change. — C

In general, when you invest money, the interest is calculated on the amount invested in the first place plus any interest already received. This is known as compound interest.

- Add the rate of interest to 100%.
- Convert this percentage to a decimal.
- Multiply the original amount by the multiplier repeatedly for each year invested.

You can also use this method to work out a repeated percentage loss or depreciation, except that to find the multiplier you must first subtract the percentage from 100%.

18.1 Percentage increase and decrease

D

1 Increase the following quantities by 10%.

a £90 **b** 50 m **c** 25 kg **d** 460 km

2 Increase the following quantities by 1%.

a £800 **b** 300 km **c** 250 *l* **d** 80 mm

3 Jo is making curtains. She needs 12 m of material.
She buys 5% more than she needs.
How much material does Jo buy?

FUNCTIONAL

4 David has a salary of £12 000 per year.
He is given a 6% pay rise.
What is his new salary?

FUNCTIONAL

5 In 2009 Llanreath Divers had 40 members.
In 2010 they had 15% more members than in 2009.
How many members did they have in 2010?

6 Decrease the following quantities by 10%.

a £70 **b** 20 m **c** 120 g **d** 48 km

7 Decrease the following quantities by 1%.

a £200 b 500 km c 350 ml d 70 cm

D

8 Shen buys a TV in a sale.
This is the price ticket on the TV.
How much does Shen pay for the TV?

HDTV
Original price £1200
Sale
20% off

FUNCTIONAL

9 A coat priced at £70 is reduced by 15% in a sale.
After 2 weeks the coat still hasn't been sold, so is reduced by a further 20%.
What is the final sale price of the coat?

The second reduction is 20% of the first sale price, not the original price.

FUNCTIONAL

D

10 Ivan gets £8 a week pocket money.
Ivan got a good report from school so his pocket money was put up by 20%.
Ivan received a detention and his pocket money was then reduced by 20%.
Is Ivan's pocket money now higher, lower or the same as before?
Explain your answer.

AO2

18.2 Calculations with money

1 Work out the VAT (15%) to be added to an MP3 player that costs £80.

FUNCTIONAL

E

2 Work out the VAT (17.5%) to be added to a mobile phone that costs £120.

FUNCTIONAL

3 Work out the total price of a guitar that costs £500 + 15% VAT.

FUNCTIONAL

D

4 Work out the total price of a pair of skis that costs £240 + 17.5% VAT.

FUNCTIONAL

5 Work out the total cost of an electricity bill that comes to £290 + 5% VAT.

FUNCTIONAL

6 Anthea buys some clothes from a catalogue.
She pays £10 the first month and then £8 per month for the next 12 months.
How much does Anthea pay in total?

FUNCTIONAL

E

7 Torben buys a racing bicycle with a cash price of £3600.
The credit terms are a deposit of 25% and 18 monthly payments of £160.

a How much is the deposit?

b What is the total of his monthly payments?

c What is the total credit price?

d How much more does it cost to buy on credit rather than pay by cash?

FUNCTIONAL

8 The cash price for a range cooker is £1350.
The credit terms for the range cooker are 20% deposit plus 24 monthly payments of £50.
What is the difference between the cash price and the credit price?

FUNCTIONAL

E

AO2

9 Carlos puts £1500 into a savings account. The interest rate is 3%.
Work out the simple interest after 5 years.

FUNCTIONAL

E

18.3 Percentage profit or loss

1 Angharad buys an antique necklace for £40 and sells it for £50.
What is her percentage profit?

2 Anil buys a flat for £82 000. He sells it five years later for £94 300.
What is his percentage profit?

3 Rowan buys wheelbarrows for £28 and sells them for £35.
Davin buys wheelbarrows for £35 and sells them for £45.
Who makes the larger percentage profit?

4 Abbie bought a step machine for £150.
She later sold it for £105.
What was her percentage loss?

5 Jim sold his drum set for £120. He had paid £160 for it.
What was his percentage loss?

6 Sacha bought a motorbike for £8250. He sold it three years later for £5280.
Helena bought a motorbike for £7850. She sold it four years later for £4867.
Who has the larger percentage loss?

18.4 Repeated percentage change

1 Garth invests £800 at rate of 3% per annum.
How much will he have at the end of two years?

2 Paulo invests £500 at a rate of 5% per annum.
How much will he have at the end of three years?

3 The population of town A is 25 000. It is increasing at a rate of 8% per annum.
The population of town B is 28 000. It is increasing at a rate of 5% per annum.
Which town will have the greater population in three years' time?

4 The population of a certain type of bat is decreasing by 10% each year.
In one colony there are 200 bats.
Estimate the number of bats in this colony in two years' time.

5 The value of a motorhome depreciates by 15% each year.
Shani buys a motorhome for £24 000.
How much will it be worth after three years?

6 In 2009 the number of students in school A is 400.
The number of students in the school is falling at a rate of approximately 5% each year.
In 2009 the number of students in school B is 490.
The number of students in the school is falling at a rate of approximately 10% each year.
Which school has the least number of students in three years' time?

Remember to round your answer to the nearest whole number; you can't have a fraction of a student!

Links to:
Foundation Student Book
Ch19, pp. 330–346

Key Points

Sequences G F E

A sequence is a set of numbers in a given order.

Each number in a sequence is called a term.

The term-to-term rule for a sequence tells you how to get from one term to the next.

Consecutive terms G F E

Terms next to each other are called consecutive terms.

To find the next term in the sequence, look at how the pattern changes between consecutive terms.

The nth (general) term F D C

The nth term of a sequence can be used to find any term in the sequence.

Finding the nth term of a linear sequence C

To find the nth term of a sequence, first look at the difference between consecutive terms.

Then compare the sequence to the multiples of the difference.

Patterns G F E C

Sequences of patterns can lead to number sequences.

Proving a theory E D C

A proof uses logical reasoning to show something is true.

Using counter-examples C

A counter-example is an example that shows a statement is incorrect.

You can disprove a statement if you can find one example that doesn't fit it.

19.1 Number sequences

1 For each sequence, first find the next three terms and then write down the term-to-term rule. G

a 10, 13, 16, 19, ...

b 22, 32, 42, 52, ...

2 For each sequence, first find the next three terms and then write down the term-to-term rule. F

a 4, 6, 9, 13, ...

b 4, 14, 34, 74, ...

3 For each sequence, first find the next three terms and then write down the term-to-term rule. E

a 50, 47, 44, 41, ...

b 50, 42, 34, 26, ...

4 Write down the first negative term in each of these sequences. E AO2

a 14, 10, 6, ...

b 50, 35, 20, ...

5 Ian has £500 in a savings account. To increase his savings he decides to pay a regular amount per month into this account. E AO3 FUNCTIONAL

The table shows the amount in his account at the end of every month.

	May	June	July	Aug
Amount in account (£)	500	650	800	950

How much will Ian have in his account at the end of January?

19.2 Using differences

E

1 For each sequence, work out the next term and describe the pattern of differences.

a 4, 5, 8, 13, ... b 12, 14, 18, 24, ...

2 Write down the next two terms in each sequence.

a 7, 5, 3, 1, ... b −1, −4, −7, −10, ...

E

3 The missing numbers from each of these sequences can be found in the cloud.

5, 7, 10, ☐, 19, ☐, 32

40, 38, ☐, 28, 20, ☐, −2

11, ☐, 14, 17, 21, 26, ☐

34 14 15 32 10 25 18 12

a Which two numbers from the cloud do **not** belong to any of the sequences?

b Write your own sequence that contains these two numbers.
Explain the pattern of differences you have used in your sequence.

4 The first four terms of a sequence are

60, 56, 48, 36, ...

What is the first term in the sequence which is less than zero?

5 A construction firm offers the following bonuses for employees who continue training while working.

Year 1	£500
Year 2	£750
Year 3	£1100
Year 4	£1550
Year 5	£2100

AO2

The scale continues in the same way.
What would the bonus be in Year 7?

19.3 Rules for sequences

E

1 The general term of a sequence is $4n$.

a Write down the first five terms of the sequence.

b Write down the tenth term.

2 Write down the first five terms of each of these sequences.

a nth term $= 2n + 1$ b nth term $= 4n - 2$ c nth term $= 8 - n$

E

AO2

3 The general term of a sequence is $2n - 10$.
Which term is the first term larger than zero?

Start by writing out the sequence one term at a time.

D

4 Write down the first five terms of each of these sequences.

a nth term $= n^2 + 1$

b nth term $= 2n^2$

C

AO2

5 The general term of a sequence is $n^2 - 10$.

a What is the 10th term of the sequence?

b Which term is the first term larger than 150?

6 The general term of a sequence is $n^2 + 25$.
Which term is the first term larger than 80?

C AO3

7 Find the nth term of each sequence.

a 6, 12, 18, 24, 30, ...
b 3, 5, 7, 9, 11, ...
c 7, 10, 13, 16, 19, ...
d 20, 15, 10, 5, ...

C

8 Find the 50th term of this sequence.
4, 10, 16, 22, 28, ...

C AO3

19.4 Sequences of patterns

1 This sequence of patterns is made from counters.

G

Pattern 1 Pattern 2 Pattern 3

a Draw patterns 4 and 5 in the sequence.

b Copy and complete the table of the number of counters needed for each pattern.

Pattern number	1	2	3	4	5
Counters	8				

2 How many counters are needed for the 10th pattern in the sequence in Q1?

F

3 Here is a pattern of dots.

E

a Without drawing the pattern, work out how many dots are in pattern 5.

b Explain how you worked out your answer to part a.

4 In a restaurant, tables can be pushed together for larger parties, as shown in the diagram.

FUNCTIONAL

C AO2

a Copy and complete the table below:

Number of tables	1	2	3	4	5
Number of people	6				

b How many people can sit at n tables?

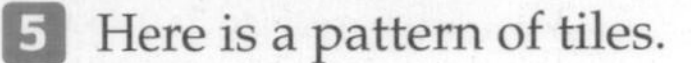

C

5 Here is a pattern of tiles.

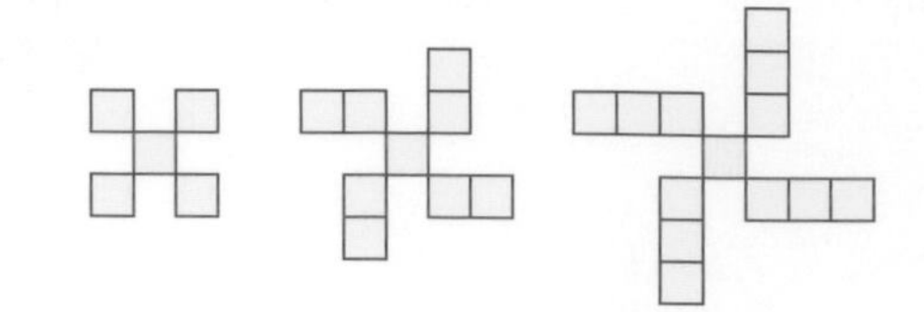

a Draw pattern 4.

b How many tiles will there be in pattern 5?

c How many tiles will there be in the nth pattern?

AO2

C

6 The manager at the restaurant in Q4 has 32 chairs altogether.
He makes a row of as many tables as he can.
How many chairs will he have left over?

AO3

19.5 Proof

E

1 The general term of a sequence is $4n$.
Show that all the terms in the sequence are multiples of 4.

AO2

E

2 I think of a positive integer, multiply it by 4 and take away 2.

a What type of number (odd or even) will I get?

b Explain how you know.

AO2

D

3 n represents an integer.

a Is $n + (n + 2)$ an odd number or an even number?

b Explain how you know.

4 n is an odd number.
Explain why $n^2 + 1$ is always an even number.

AO2

19.6 Using counter-examples

C

1 Tina says, 'When you multiply an integer by a decimal, the answer is always smaller than the integer you started with.'
Give a counter-example to show that Tina is wrong.

2 Stephen says, 'When you square a number, the answer is always larger than the number you started with.'
Give a counter-example to show that Stephen is wrong.

3 Inca says, '$n^2 + 3$ is never a multiple of 7.'
Give a counter-example to show that Inca is wrong.

4 Paddy says, 'If a and b are both prime numbers, $a + b$ cannot be an odd number.'
Give a counter-example to show that Paddy is wrong.

C

5 Claire says, 'If x is an odd integer, then $x^2 - 1$ is always odd.'
Explain why Claire is wrong.

AO2

Coordinates and linear graphs

Links to:
Foundation Student Book
Ch20, pp. 347–368

Key Points

Coordinates of a point (G F)

The coordinates of a point tell you its position on a grid.

The first value gives the number of units left or right in the x-direction.

The second number gives the number of units up or down in the y-direction.

Conversion graphs (F E)

A conversion graph converts values from one unit to another.

Conversion graphs can be linear or curved.

Straight-line graphs (E)

A straight-line graph parallel to the x-axis has equation $y =$ a number.

A straight-line graph parallel to the y-axis has equation $x =$ a number.

Straight-line graphs have equations of the form $y = mx + c$.

Graphs of functions (E D)

You can draw a graph of the function $y = 2x + 1$.

Substitute values for x

Write the values of x and the corresponding values of y in a table.

Plot the (x, y) coordinate pairs on a grid.

Join the points with a straight line.

Distance–time graphs (E D C)

A distance–time graph represents a journey. The x-axis (horizontal) represents the time taken. The y-axis (vertical) represents the distance from the starting point.

Mid-point of a line segment (D C)

$$\text{mid-point } (x, y) = \left(\frac{x_1 + x_2}{2}, \frac{y_1 + y_2}{2}\right)$$

where (x_1, y_1) and (x_2, y_2) are the coordinates of the end-points.

Using the equation $y = mx + c$ (C)

A straight-line graph can be described by the equation $y = mx + c$.

The gradient (slope), m, of a straight line measures how steep it is.

$$\text{Gradient, } m = \frac{\text{change in } y}{\text{change in } x}$$

If the line slopes upward, the gradient is positive.

If the line slopes downward, the gradient is negative.

Parallel lines have the same gradient.

c is the y-intercept

Rates of change (C)

A straight-line graph shows that the rate of change is steady.

A curved graph shows that the rate of change varies.

The steeper the line, the faster the rate of change.

20.1 Coordinates and line segments

G

1 Write down the coordinates of the points A, B, C and D.

F

2 EFGH is a rectangle.

a Write down the coordinates of the points E, F, and G.

b Copy the diagram.

c Draw the point H to complete the rectangle.

d Write down the coordinates of the point H.

A02

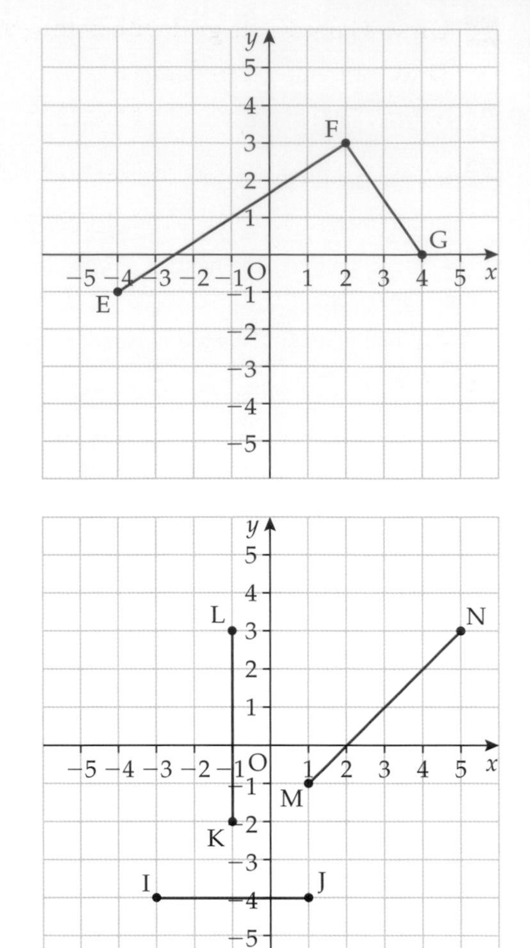

D

3 For each line segment:

a write down the coordinates of the end-points

b work out the coordinates of the mid-point.

C

4 Work out the mid-points of these line segments:

a PQ: P(−2, 4) and Q(2, 4)

b RS: R(5, 1) and S(5, −2)

c TU: T(−3, 1) and U(1, −3)

C

A03

5 The point V has coordinates (2, 2).
The mid-point of the line segment VW has coordinates (2, −1).
Work out the coordinates of the point W.

20.2 Plotting straight-line graphs

E

1 Write down the equations of the lines A and B.

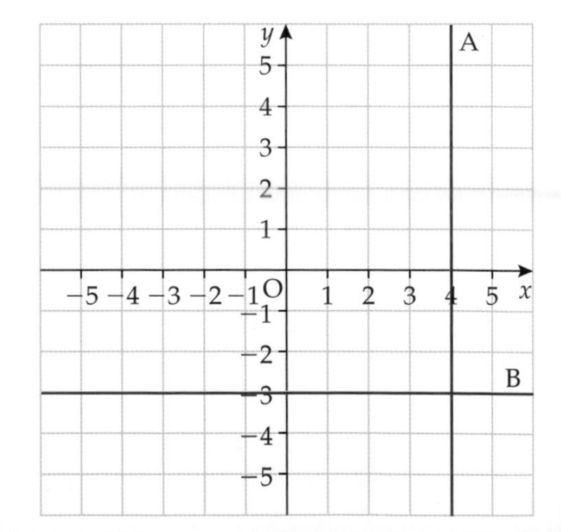

2 Draw a coordinate grid with both x- and y-axes from -4 to $+4$.
Draw and label these graphs:

a $y = 2$ b $x = -3$

3 a Copy and complete this table of values for $y = x + 3$.

x	-2	-1	0	1	2
y	1				

b Draw the graph of $y = x + 3$.

4 a Copy and complete this table of values for $y = 2x - 1$.

x	-2	-1	0	1	2
y	-5				

b Draw the graph of $y = 2x - 1$.

c Draw the line $x = -1$ on your graph.

d Z is the point where the two lines cross.
Mark the point Z and write down its coordinates.

5 Simon says, 'The lines $y = 3x + 1$ and $x = 2$ cross at the point $(2, 5)$.'
Is Simon correct? Show working to support your answer.

E

D

AO2

D

AO3

20.3 Gradients of straight-line graphs

1 a Draw a coordinate grid with both x- and y-axes from 0 to 10.
On the same grid, draw the graphs of
i $y = 2x$ ii $y = x + 3$ iii $y = x + 1$

b Which line is the steepest?

c How can you tell which line is steepest from the equations?

d Which lines are parallel to each other?

e How can you tell which lines are parallel from the equations?

D

AO2

2 Work out the gradients of lines A and B.

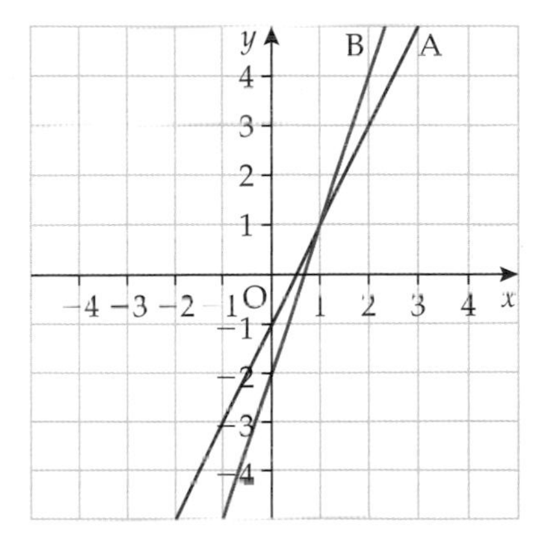

C

3 Write down the gradient of these straight lines

a $y = 4x + 1$ b $y = 7 + 2x$ c $y = \frac{1}{4}x - 3$

4 Which of these straight lines are parallel to $y = 2x + 8$?

a $y = 3x + 8$ b $y = 1 + 2x$ c $y = 2x - 2$

C

5 Write down the y-intercepts of these lines:

a $y = 4x + 1$ b $y = 7 + 2x$ c $y = \frac{1}{4}x - 3$

6 Write the following equations in the form $y = mx + c$.

a $2y = 6x + 10$ b $y - 3 = 3x$ c $5y = -10x + 25$

C

7 Without plotting these straight lines, identify the ones parallel to the line $y = 2x + 3$.

a $3y = 6x - 9$ b $y = 3 - 2x$ c $y - 2x = 5$

d $4y = 12x + 12$ e $y + 2x = 8$ f $2y = 4x + 14$

AO2

20.4 Conversion graphs

F

1 This is a conversion graph between pounds sterling (£) and American dollars ($).

Use the graph to find the value of

a £20 in dollars

b $64 in pounds

c £44 in dollars.

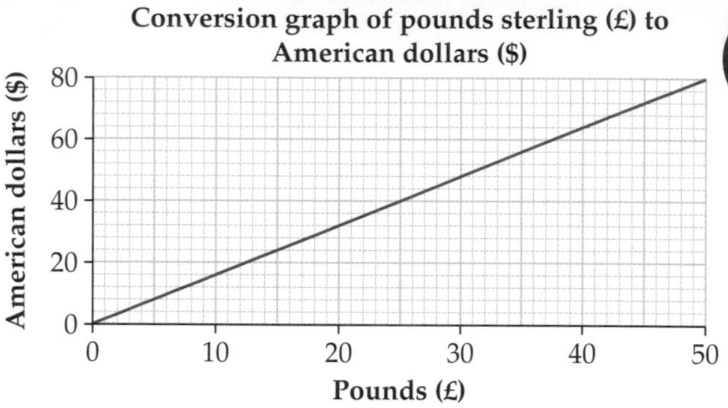

2 This is a conversion graph between pounds sterling (£) and Polish zloty (Zl).

a Use the graph to find the value of

i £5 in Polish zloty

ii 10 Zl in pounds

b Use your answers to part **a** to find the value of

i £50 in Polish zloty

ii 30 Zl in pounds.

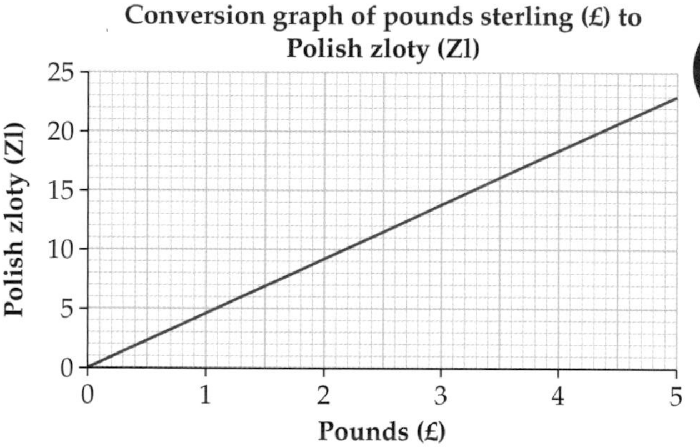

AO2

E

3 The conversion rate from pounds sterling (£) to South African rand (R) is £1 = 12 R.

FUNCTIONAL

a Copy and complete this conversion table between pounds and rand.

£ (x)	0	5	10
R (y)		60	

b Draw a conversion graph with x-values from £0 to £10 and y-values from 0 R to 120 R.

c Use your graph to convert:

i £4 to rand

ii 90 R to pounds.

d Use your answers to part c to find the value of

i £32 in rand

ii 270 R in pounds.

AO2

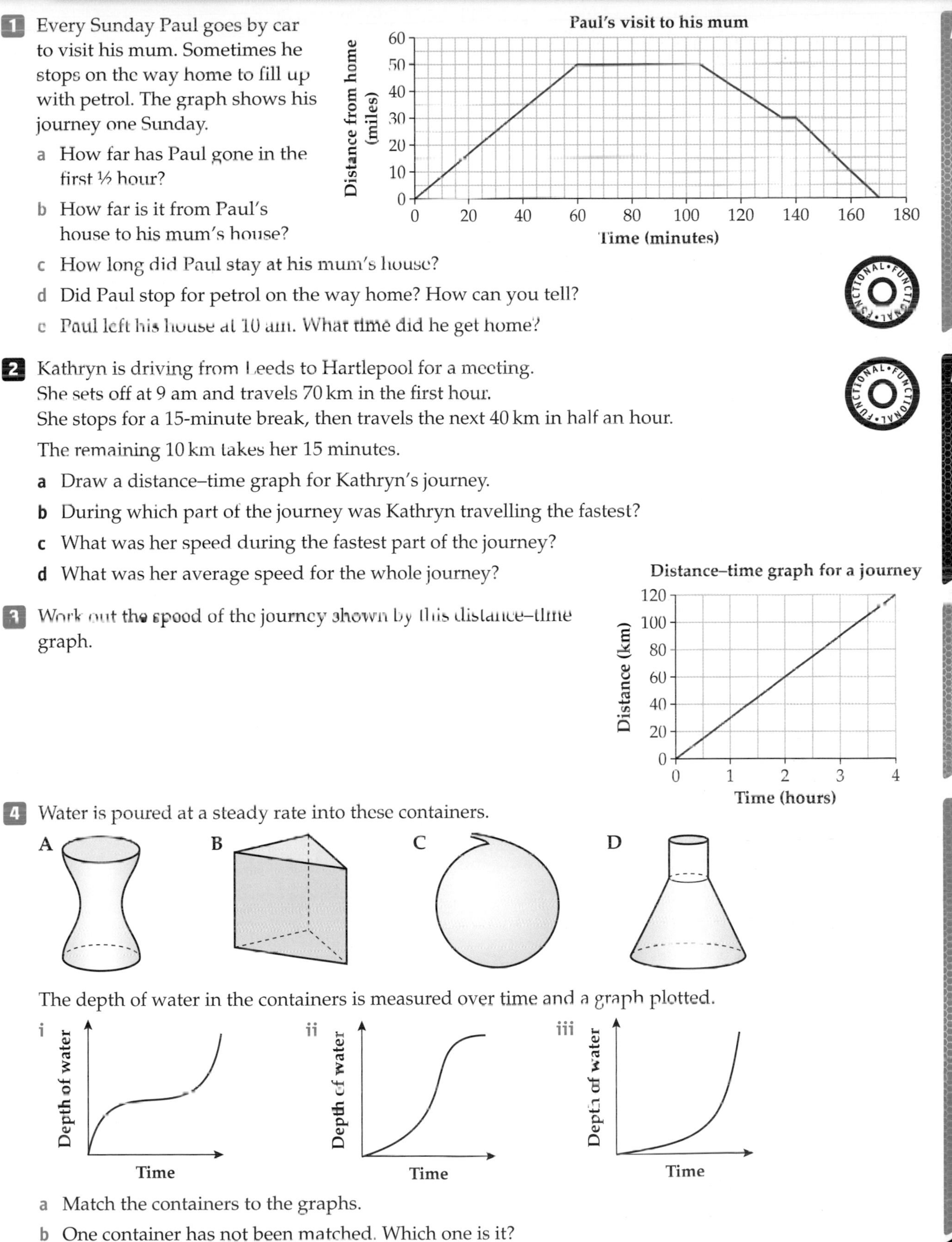

1 Every Sunday Paul goes by car to visit his mum. Sometimes he stops on the way home to fill up with petrol. The graph shows his journey one Sunday.

a How far has Paul gone in the first ½ hour?

b How far is it from Paul's house to his mum's house?

c How long did Paul stay at his mum's house?

d Did Paul stop for petrol on the way home? How can you tell?

e Paul left his house at 10 am. What time did he get home?

2 Kathryn is driving from Leeds to Hartlepool for a meeting.
She sets off at 9 am and travels 70 km in the first hour.
She stops for a 15-minute break, then travels the next 40 km in half an hour.

The remaining 10 km takes her 15 minutes.

a Draw a distance–time graph for Kathryn's journey.

b During which part of the journey was Kathryn travelling the fastest?

c What was her speed during the fastest part of the journey?

d What was her average speed for the whole journey?

3 Work out the speed of the journey shown by this distance–time graph.

4 Water is poured at a steady rate into these containers.

The depth of water in the containers is measured over time and a graph plotted.

i ii iii

a Match the containers to the graphs.

b One container has not been matched. Which one is it?

c Sketch a graph for this container.

Number skills revisited

Links to:
Foundation Student Book
Ch21, pp. 371–375

This section is revising the number skills that you will need to use in Unit 3.

1 a Use the numbers from the cloud to complete these equivalent fractions.

i $\frac{2}{3} = \frac{\square}{6}$ ii $\frac{3}{4} = \frac{9}{\square}$ iii $\frac{\square}{5} = \frac{12}{30}$ iv $\frac{6}{\square} = \frac{18}{21}$

2 5 4 12 7

b What number from the cloud haven't you used?

c Write down the reciprocal of this number.

2 At a football match there were 2000 supporters altogether.
The table shows how many of the supporters were men, women, boys and girls.

	Men	Women	Boys	Girls
Number of supporters	840	420	600	140

a What percentage of the supporters were:

i men ii women iii boys iv girls?

b Write the ratio of men : women in its simplest form.

c Write the ratio of girls : women in its simplest form.

3 Copy and complete the table.
Write the fractions in their simplest form.

Percentage	20%				15%
Decimal		0.7		0.05	
Fraction			$\frac{3}{10}$		

4 Silvana has these two number cards.

18 25

a Work out the sum of the two numbers.

b Work out the difference of the two numbers.

c Work out the product of the two numbers.

5 Work out the missing numbers in each of these calculations.

a $5 \times 2 + 6 = \square$ b $12 - 3 \times 4 = \square$ c $(10 + 6) \div 4 = \square$

d $9 + 3 \times \square = 18$ e $8 \times 2 - \square = 5$ f $(\square + 2) \times 6 = 30$

g $5^2 + 4 = \square$ h $7^2 - \square = 40$ i $\square \times 3^2 = 54$

6 Work out

a 40×0.7 b 36×0.3 c 4.2×2.5

d $64 \div 0.2$ e $15.9 \div 3$ f $9.75 \div 0.15$

7 Write whether A, B or C is the correct answer for each of these.

a 2.46 rounded to 1 d.p. is A 2.4 B 2.5 C 2.0

b 13.664 rounded to 2 d.p. is A 13.7 B 13.66 C 13.67

c 0.0956 rounded to 3 d.p. is A 0.956 B 0.095 C 0.096

d 345 rounded to 1 s.f. is A 3 B 350 C 300

e 489 rounded to 2 s.f. is A 490 B 49 C 500

f 0.6679 rounded to 3 s.f. is A 0.668 B 0.67 C 0.667

8 a What is the reciprocal of $\frac{2}{7}$?

b Multiply $\frac{2}{7}$ by its reciprocal. What result do you get?

Links to:
Foundation Student Book
Ch22, pp. 376–396

Key Points

Angle properties

F E D

Angles on a straight line add up to 180°.

$a + b + c = 180°$

Angles around a point add up to 360°.

$d + e + f + g + h = 360°$

Vertically opposite angles are equal.

$a = b, c = d$

Corresponding angles are equal.

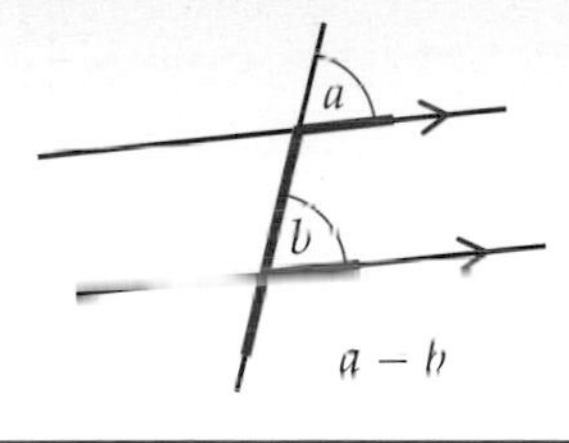

Alternate angles are equal.

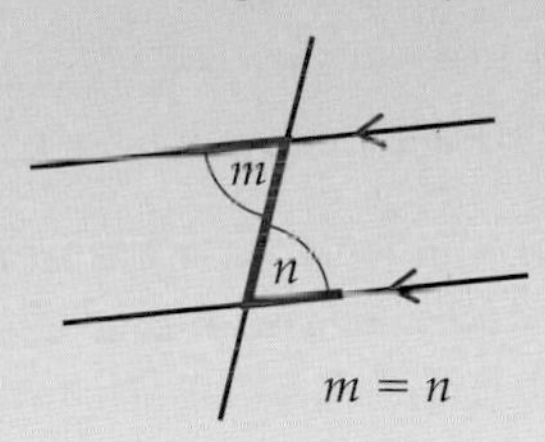

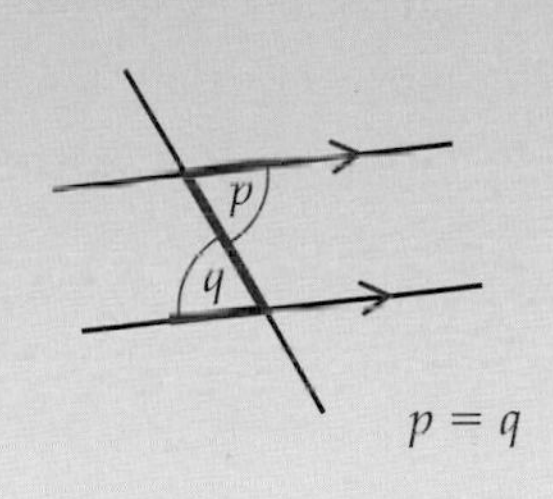

Bearings

E D C

A bearing gives a direction in degrees.
It is always measured clockwise from north.

It can have any value from 0° to 360°. It is always written with three figures.

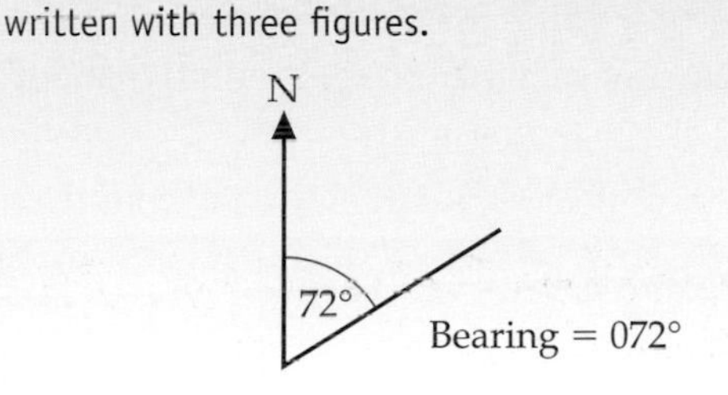

22.1 Angles and turning

1 Here are five description cards and four answer cards.

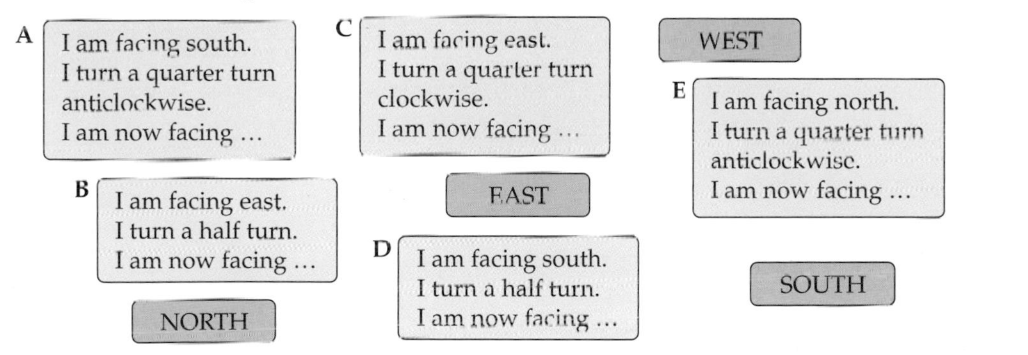

a Match each answer card to the correct description card.

b You need one more answer card.
What direction must be on the missing answer card?

G

2 Suzie starts facing west. She turns to face south.

a In which direction did she turn?

b What was the size of the turn?

c Is there another turn that Suzie could have made to end up facing south?

G

AO2

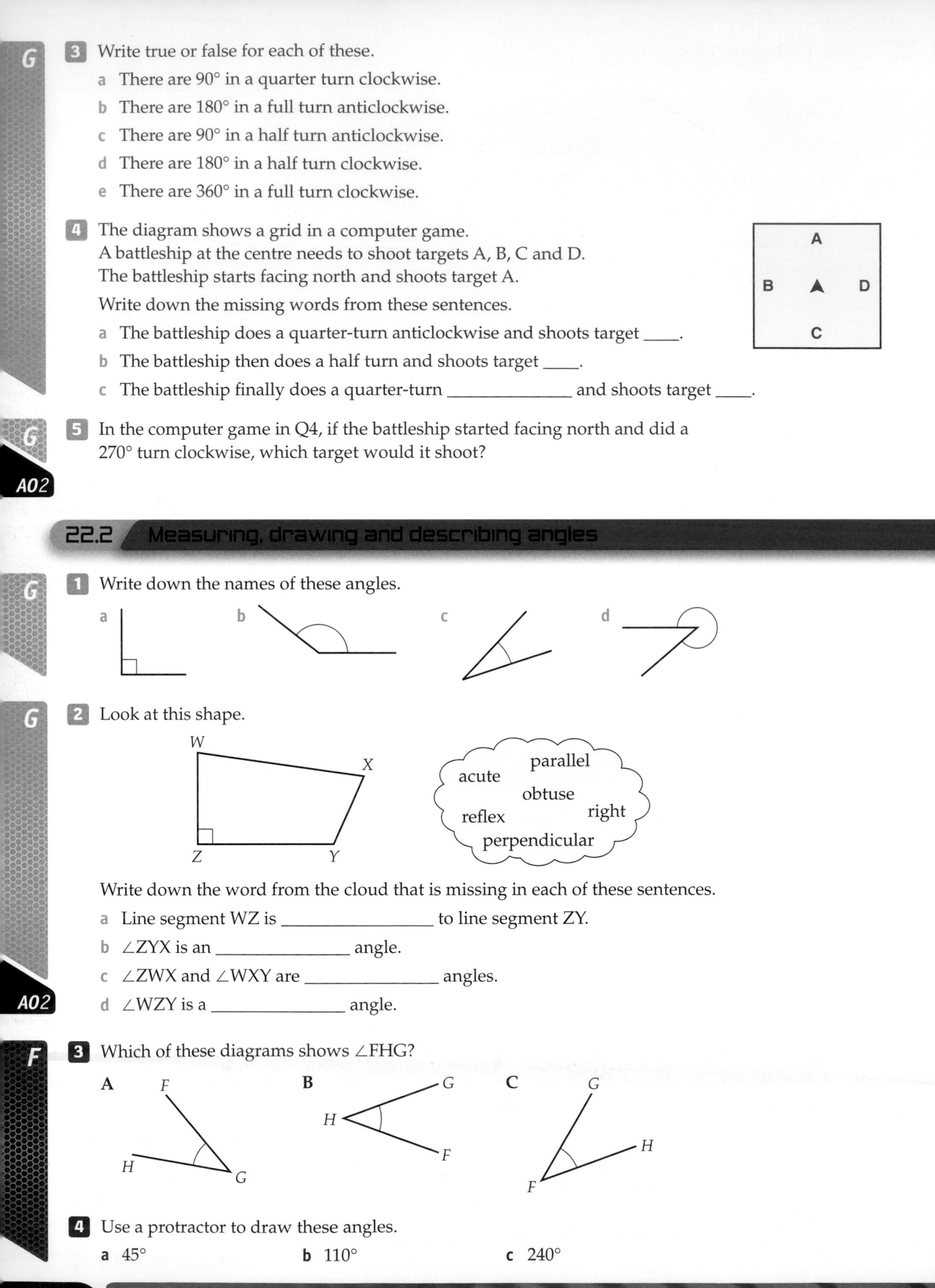

G

3 Write true or false for each of these.

a There are 90° in a quarter turn clockwise.

b There are 180° in a full turn anticlockwise.

c There are 90° in a half turn anticlockwise.

d There are 180° in a half turn clockwise.

e There are 360° in a full turn clockwise.

4 The diagram shows a grid in a computer game.
A battleship at the centre needs to shoot targets A, B, C and D.
The battleship starts facing north and shoots target A.

Write down the missing words from these sentences.

a The battleship does a quarter-turn anticlockwise and shoots target ____.

b The battleship then does a half turn and shoots target ____.

c The battleship finally does a quarter-turn ____________ and shoots target ____.

G AO2

5 In the computer game in Q4, if the battleship started facing north and did a 270° turn clockwise, which target would it shoot?

22.2 Measuring, drawing and describing angles

G

1 Write down the names of these angles.

a b c d

G AO2

2 Look at this shape.

Write down the word from the cloud that is missing in each of these sentences.

a Line segment WZ is ______________ to line segment ZY.

b ∠ZYX is an ______________ angle.

c ∠ZWX and ∠WXY are ______________ angles.

d ∠WZY is a ______________ angle.

F

3 Which of these diagrams shows ∠FHG?

4 Use a protractor to draw these angles.

a 45° **b** 110° **c** 240°

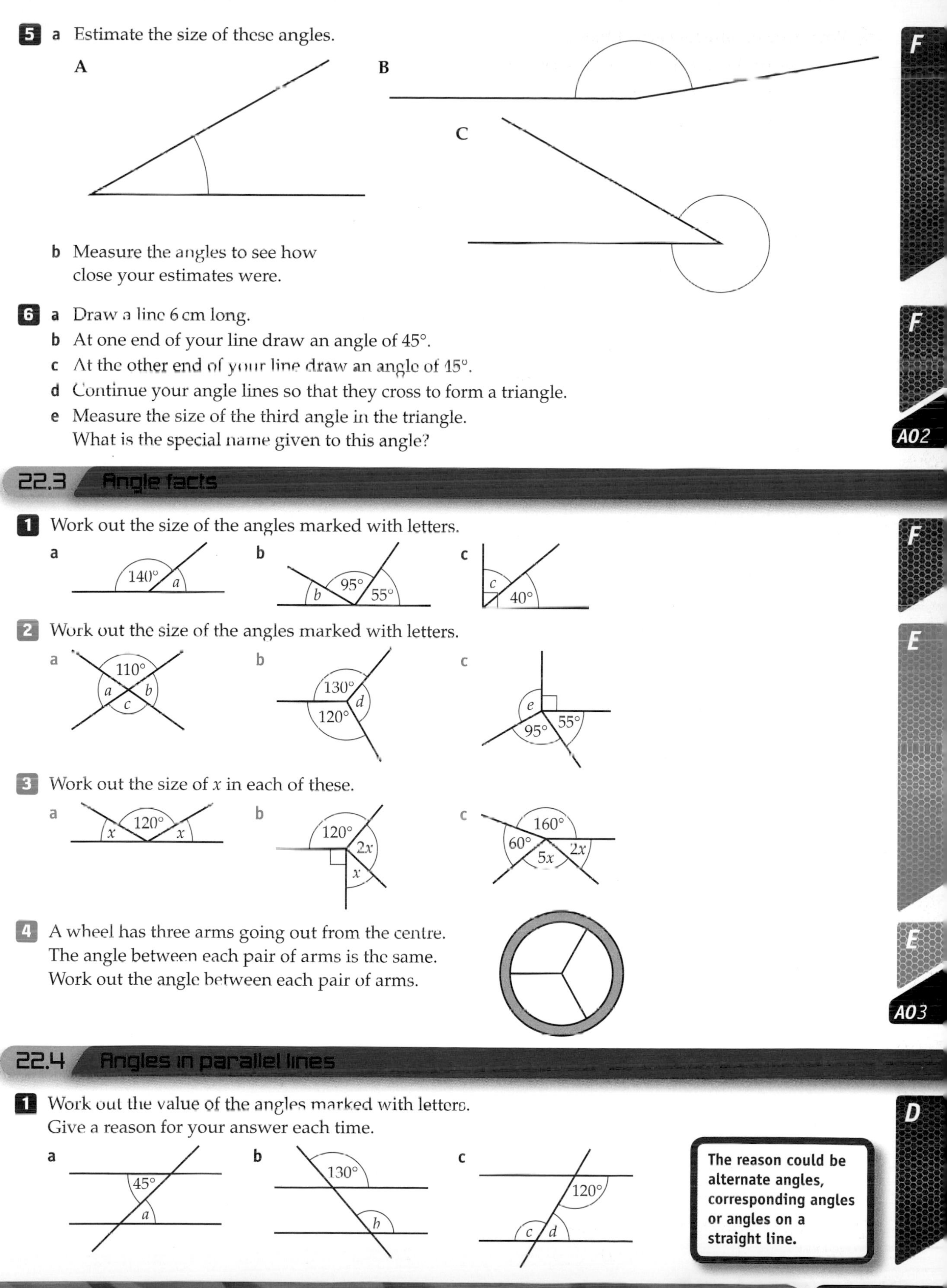

5 a Estimate the size of these angles.

A B C

b Measure the angles to see how close your estimates were.

6 a Draw a line 6 cm long.
b At one end of your line draw an angle of 45°.
c At the other end of your line draw an angle of 15°.
d Continue your angle lines so that they cross to form a triangle.
e Measure the size of the third angle in the triangle. What is the special name given to this angle?

F F AO2

22.3 Angle facts

1 Work out the size of the angles marked with letters.

a $140°$, a b b, $95°$, $55°$ c c, $40°$

2 Work out the size of the angles marked with letters.

a $110°$, a, b, c b $130°$, d, $120°$ c e, $95°$, $55°$

3 Work out the size of x in each of these.

a x, $120°$, x b $120°$, $2x$, x c $160°$, $60°$, $5x$, $2x$

4 A wheel has three arms going out from the centre.
The angle between each pair of arms is the same.
Work out the angle between each pair of arms.

F E E AO3

22.4 Angles in parallel lines

1 Work out the value of the angles marked with letters.
Give a reason for your answer each time.

a $45°$, a b $130°$, b c $120°$, c, d

The reason could be alternate angles, corresponding angles or angles on a straight line.

D

D

2 Work out the value of the angles marked with letters.
Give a reason for your answer each time.

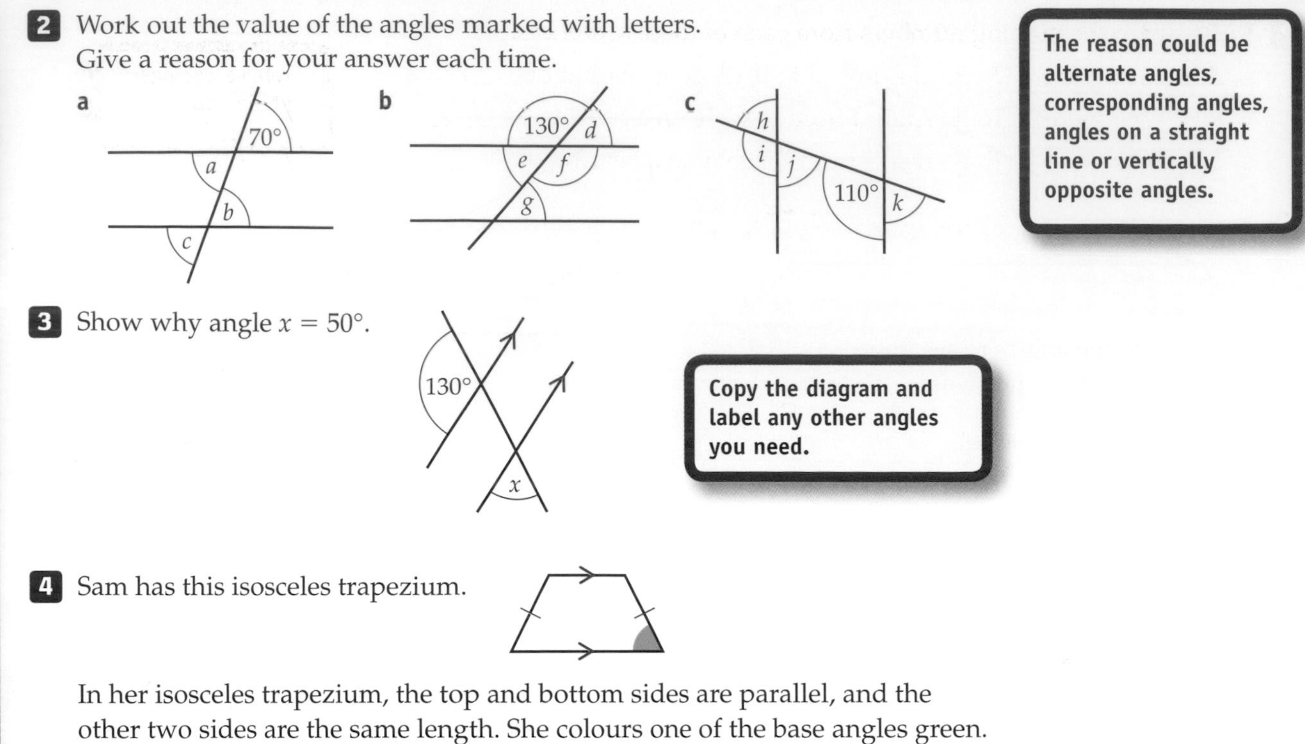

> **The reason could be alternate angles, corresponding angles, angles on a straight line or vertically opposite angles.**

D

3 Show why angle $x = 50°$.

> **Copy the diagram and label any other angles you need.**

4 Sam has this isosceles trapezium.

In her isosceles trapezium, the top and bottom sides are parallel, and the other two sides are the same length. She colours one of the base angles green.
Sam draws this tessellation using her isosceles trapezium.

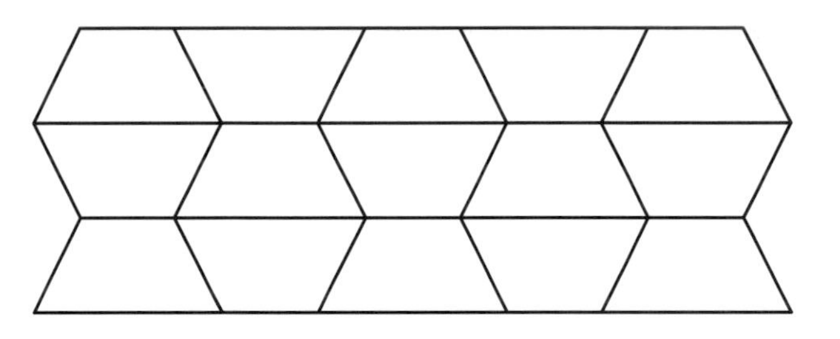

A03

Sam colours in all of the angles that are the same size as the green angle.
How many angles does she colour green?

22.5 Bearings

E

1 For each diagram measure the bearing of X from Y.

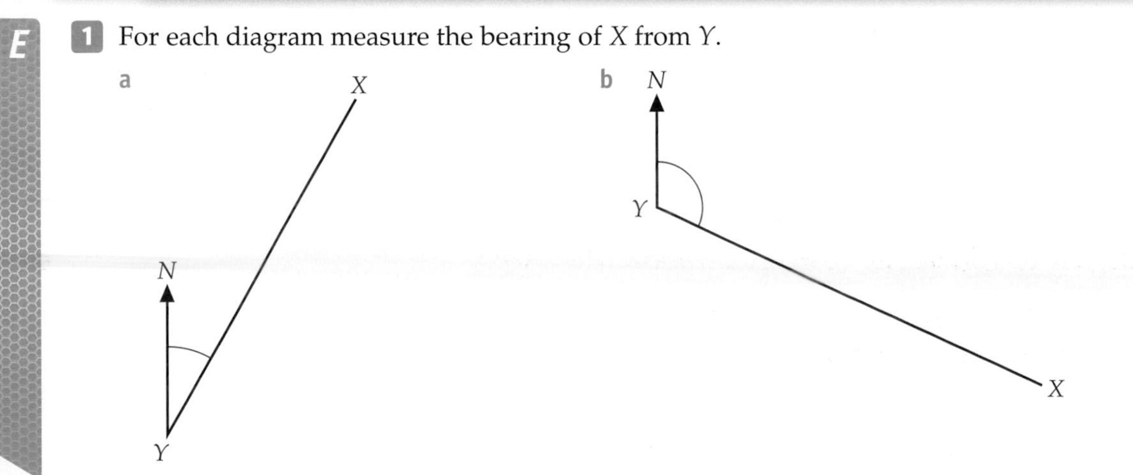

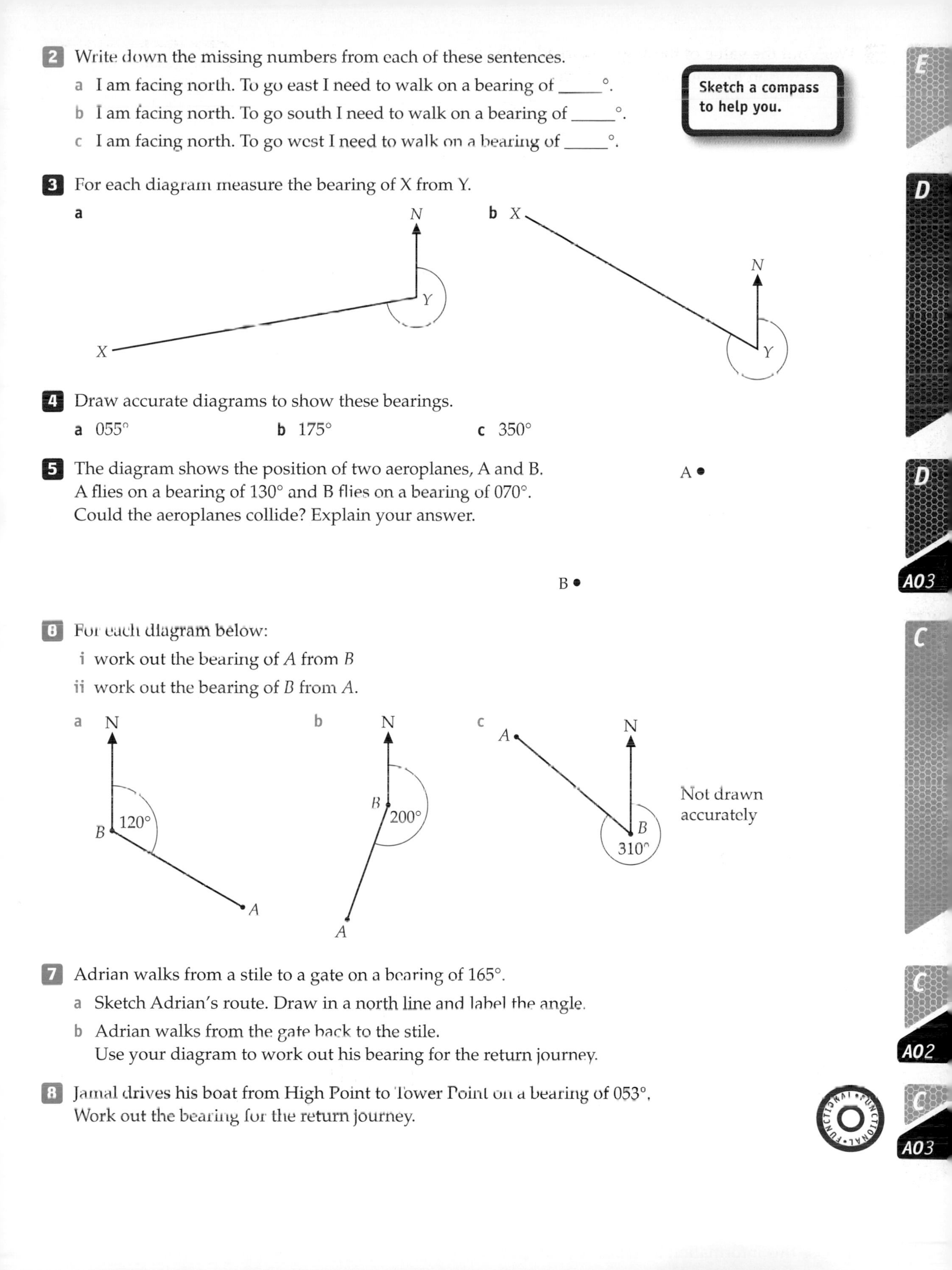

2 Write down the missing numbers from each of these sentences.

a I am facing north. To go east I need to walk on a bearing of _____°.

b I am facing north. To go south I need to walk on a bearing of _____°.

c I am facing north. To go west I need to walk on a bearing of _____°.

Sketch a compass to help you.

3 For each diagram measure the bearing of X from Y.

4 Draw accurate diagrams to show these bearings.

a 055° **b** 175° **c** 350°

5 The diagram shows the position of two aeroplanes, A and B.
A flies on a bearing of 130° and B flies on a bearing of 070°.
Could the aeroplanes collide? Explain your answer.

6 For each diagram below:

i work out the bearing of *A* from *B*

ii work out the bearing of *B* from *A*.

7 Adrian walks from a stile to a gate on a bearing of 165°.

a Sketch Adrian's route. Draw in a north line and label the angle.

b Adrian walks from the gate back to the stile.
Use your diagram to work out his bearing for the return journey.

8 Jamal drives his boat from High Point to Tower Point on a bearing of 053°.
Work out the bearing for the return journey.

E D D AO3 C C AO2 C AO3 FUNCTIONAL

23 Measurement 1

Links to:
Foundation Student Book
Ch23, pp. 397–408

Key Points

Reading scales G

To read a scale you need to work out what each division represents.

Estimating G F

You can use standard metric units to estimate length (or distance), volume (or capacity) and mass. Choose the units you use carefully, and check your estimates by comparing them to measurements that you know.

Times and dates G F

The 12-hour clock uses am for the morning and pm for the afternoon.

The 24-hour clock counts a whole day from 0000 to 2359.

There are 12 months in a year.

Time and timetables F E

When you are solving problems involving time, be careful converting between hours and minutes.

Remember: 1 day = 24 hours; 1 hour = 60 minutes;
1 minute = 60 seconds

23.1 Estimating

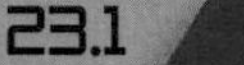

G

1 What units would you use to measure each of these?
Choose your answer from the cards on the right.

a the height of a building
b the mass of a dog
c the length of a drawing pin
d the capacity of a mug
e the distance from London to Edinburgh
f the mass of an elephant
g the capacity of a bucket
h the mass of an apple

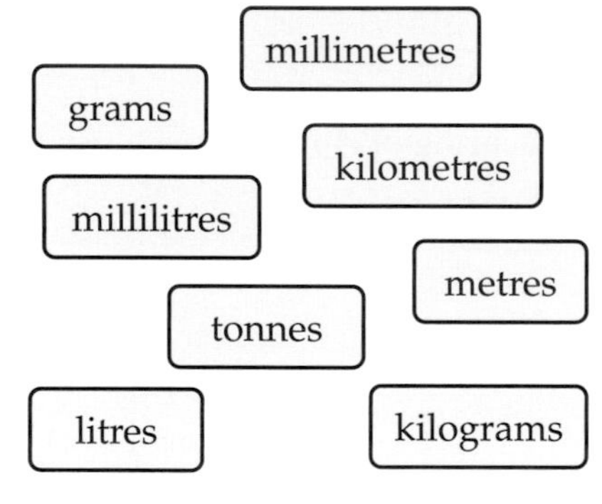

F

2 Which of these estimates are sensible? Give a better estimate where necessary.

a The length of a calculator is 15 cm.
b The mass of horse is 50 kg.
c The distance between two towns is 200 mm.
d The capacity of a bottle of lemonade is 1.5 *l*.
e The mass of a man is 75 kg.

F

3 **a** Estimate the width of your TV at home.
b Measure the width of your TV with a ruler.
How close was your estimate?

4 **a** Estimate the width of your calculator.
b Measure the width of your thumb using a ruler.
c How many thumb-widths can you fit across your calculator?
d Use this information to improve your estimate.

A02

23.2 Reading scales

1 The diagram shows a tyre pressure gauge.
Tyre pressures are measured in psi (pounds per square inch).

a What does each small division on this scale represent?

b What is the reading on the scale?

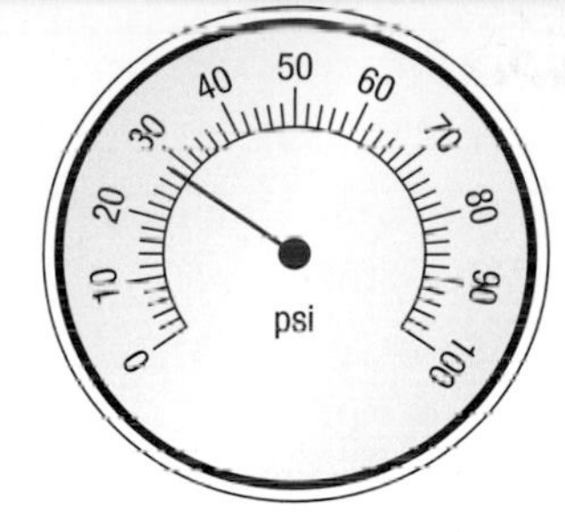

2 What are the values shown on each of these scales?

a

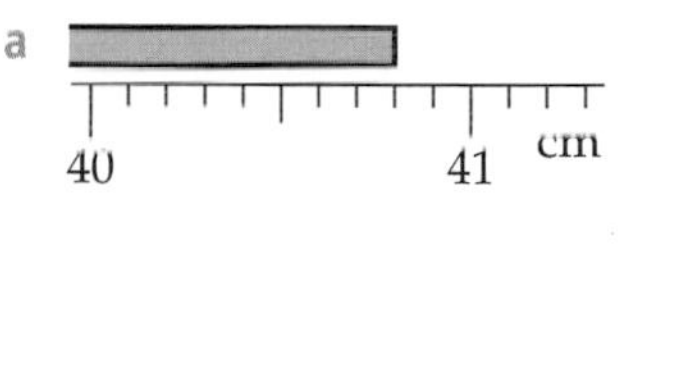

b

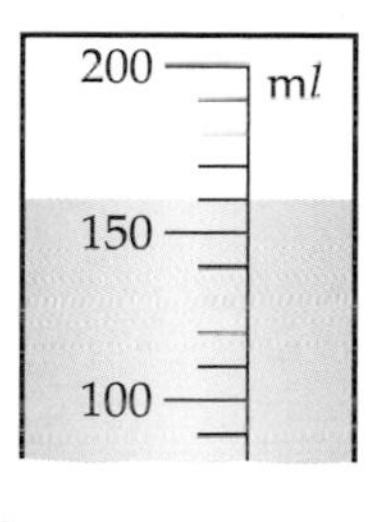

c

3 This diagram shows some water in a measuring jug.

a Write down the number of m*l* that arrow A is pointing to.

b Write down the number of m*l* that arrow B is pointing to.

c What is the difference between the two measurements?

d Carly pours 50 m*l* of water out of the jug. How much water is left?

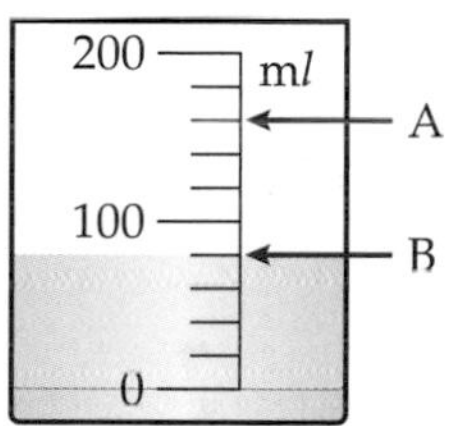
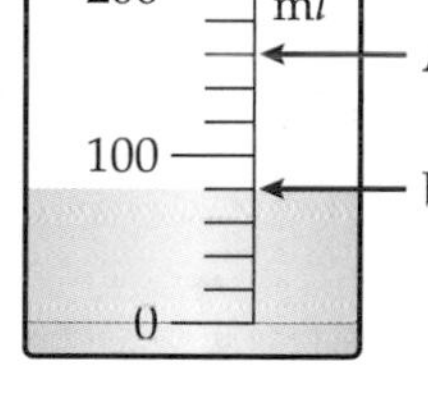

4 Estimate the reading on each of these scales.

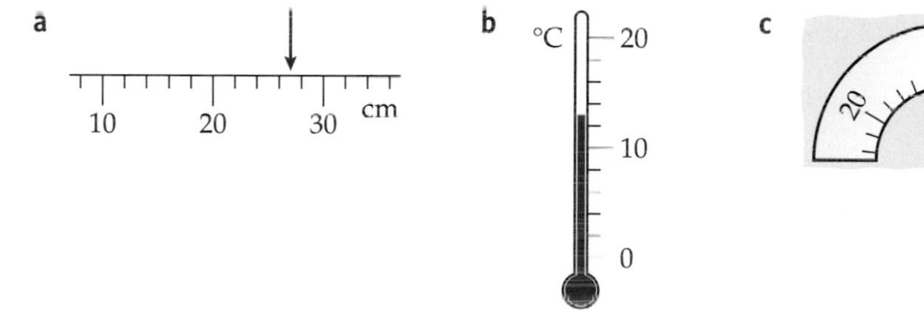

c

5 Alun is driving his car. The diagram below shows the speed he is travelling.
Alun says 'I am travelling at nearly 121 km/h.'
Is he correct?
Explain your answer fully.

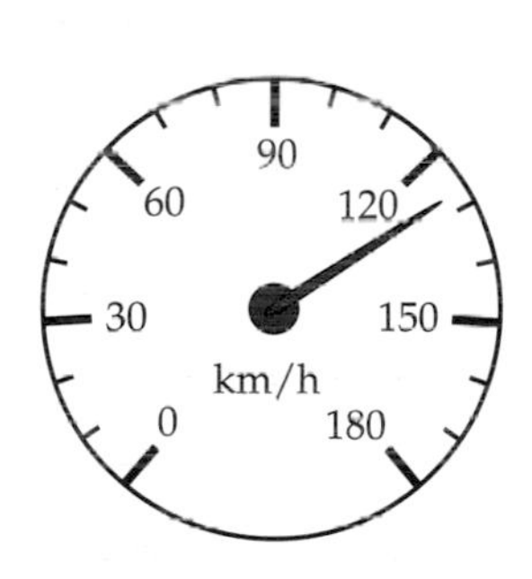

23.3 Times and dates

1 Write down the time this digital clock is showing using the 12-hour clock.

18:30

2 Write down two different times that this clock could be showing using the 24-hour clock.

F

3 Write true or false for each of these.

a March has 31 days.

b June has 30 days.

c September has 31 days.

d If today is Thursday 13th October, exactly one week ago the date was Thursday 7th October.

e If today is Monday 20th March, in exactly two weeks' time, the date will be Monday 3rd April.

f If today is Saturday 28th June, in exactly three weeks' time, the date will be Saturday 18th July.

F

4 The diary shows the Year 10 June exam timetable for Netherwood school.

FUNCTIONAL • FUNCTIONAL •

Monday	Tuesday	Wednesday	Thursday	Friday
	1	2	3	4
7 am History pm Maths 1	8 am Science 1 pm English 1	9 am Geography pm French	10 am RE pm Science 2	11 am Graphics pm English 2
14 am Art 1 pm Art 2	15 am DT pm Media	16 am German pm Spanish	17	18 am ICT pm Maths 2
21	22 am Textiles pm Drama	23	24 am Music pm PE	25 am Geology
28	29	30		

a On which day of the week are there two language exams?

b On which date are there only art exams?

c On which date is there a science exam in the afternoon?

d How many nights of revision do students have between the two maths exams?

e Teachers have been asked to tell students their results exactly one week after the final exam. What will be the date when the students get their results?

AO2

f The Year 10 trip is planned for the second Monday in July. What is the date of the Year 10 trip?

23.4 Time and timetables

F

1 Write whether A, B or C is the correct answer for each of these.

a $2\frac{1}{2}$ hours is the same as

A 90 minutes B 120 minutes C 150 minutes.

b $3\frac{1}{4}$ hours is the same as

A 195 minutes B 180 minutes C 135 minutes.

c $\frac{1}{6}$ of an hour is the same as

A 6 minutes B 10 minutes C 36 minutes.

d 45 minutes is the same as

A 0.45 hours B $\frac{9}{20}$ of an hour C $\frac{3}{4}$ of an hour

e 20 minutes is the same as

A $\frac{1}{3}$ of an hour B 0.2 hours C $\frac{1}{20}$ of an hour

2 Work out the period of time between

a 10.30 and 15:45 b 09:30 and 13:05.

3 Caroline works at a children's nursery.
This time sheet shows the hours she works one week.

Day	Start time	Finish time	Time worked
Monday	09:00	12:30	3 hours 30 mins
Tuesday	08:30	12:45	
Wednesday	08:45	14:00	
Thursday	09:15	15:00	
Friday	07:30	16:45	

a Copy and complete Caroline's timesheet.

b Work out the total time that Caroline worked during this week.

c Caroline is paid £8 per hour.
How much did Caroline earn this week?

4 The train journey from Leeds to Newcastle should take 1 hour and 40 minutes.
Kalpana's train leaves Leeds at 9:05 am.
What is the arrival time of the train into Newcastle?

5 The timetable shows the times of evening trains travelling between Swansea and Cardiff.

Swansea – Cardiff Railway Timetable

Swansea	1910	—	1929	1955	—	2055
Llansamlet	1917	—	—	—	—	—
Skewen	1921	—	—	2001	—	—
Neath	1925	—	1940	2006	—	2106
Port Talbot	1936	—	1948	2013	—	2113
Pyle	1941	—	—	—	—	—
Bridgend	1950	—	1959	2025	2043	2125
Pencoed	—	—	—	—	2049	—
Cardiff Central	2010	—	2022	2047	2112	2147

a How long does a journey from Port Talbot to Cardiff Central take?

b Ellen arrives at Port Talbot train station at 1952. She wants to catch a train to Cardiff. How long does she have to wait for the next train?

c Berwyn wants to travel from Swansea to Skewen.
He arrives at Swansea train station at 1925.
How long does he have to wait for the next train to go to Skewen?

Links to:
Foundation Student Book
Ch24, pp. 409–420

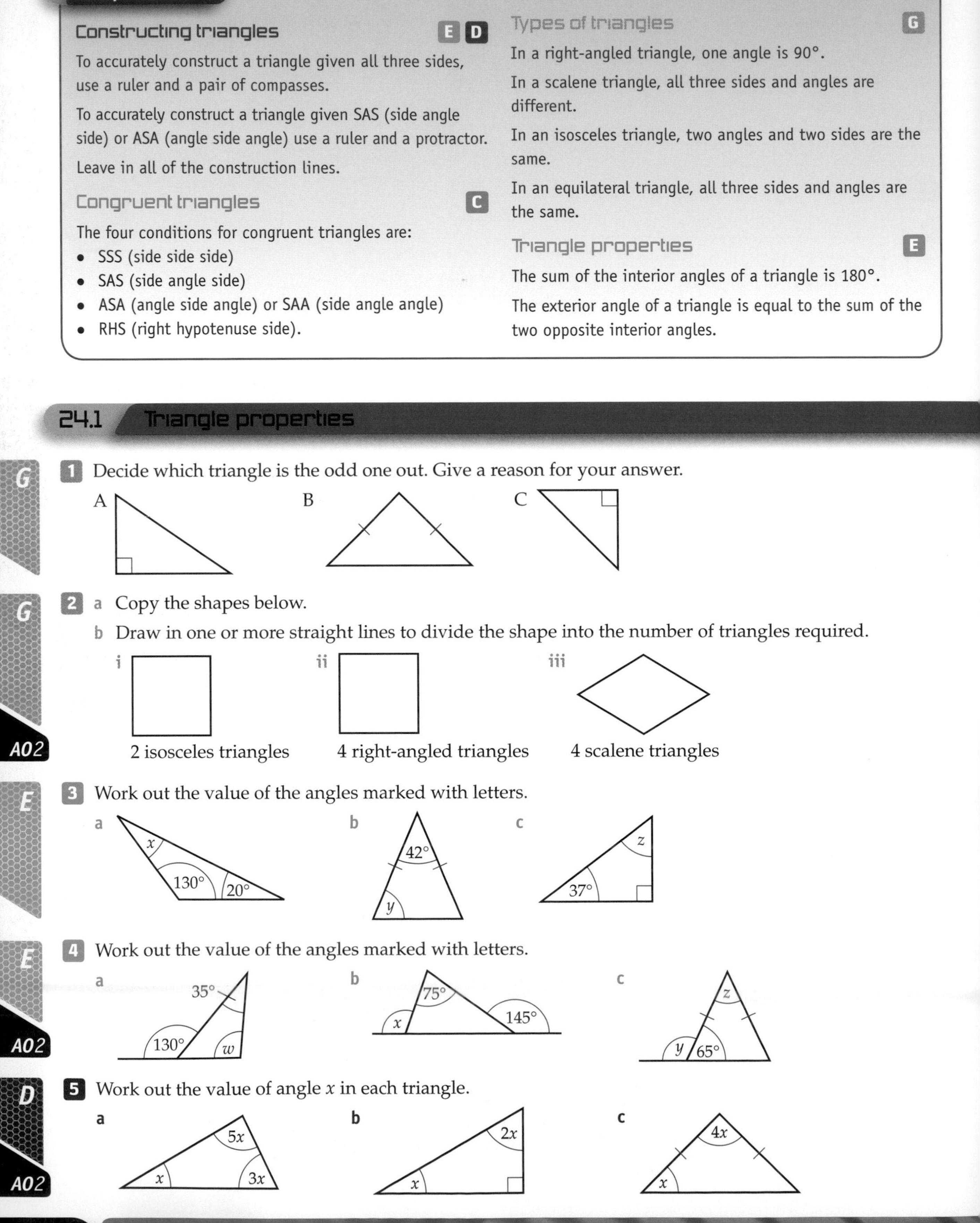

Key Points

Constructing triangles (E D)

To accurately construct a triangle given all three sides, use a ruler and a pair of compasses.

To accurately construct a triangle given SAS (side angle side) or ASA (angle side angle) use a ruler and a protractor.

Leave in all of the construction lines.

Congruent triangles (C)

The four conditions for congruent triangles are:

- SSS (side side side)
- SAS (side angle side)
- ASA (angle side angle) or SAA (side angle angle)
- RHS (right hypotenuse side).

Types of triangles (G)

In a right-angled triangle, one angle is 90°.

In a scalene triangle, all three sides and angles are different.

In an isosceles triangle, two angles and two sides are the same.

In an equilateral triangle, all three sides and angles are the same.

Triangle properties (E)

The sum of the interior angles of a triangle is 180°.

The exterior angle of a triangle is equal to the sum of the two opposite interior angles.

24.1 Triangle properties

G

1 Decide which triangle is the odd one out. Give a reason for your answer.

G / AO2

2 a Copy the shapes below.

b Draw in one or more straight lines to divide the shape into the number of triangles required.

E

3 Work out the value of the angles marked with letters.

E / AO2

4 Work out the value of the angles marked with letters.

D / AO2

5 Work out the value of angle x in each triangle.

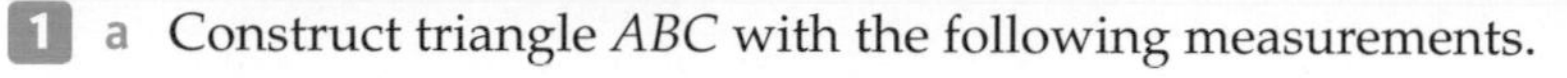

24.2 Constructing triangles

1 a Construct triangle ABC with the following measurements.
$AB = 8\,\text{cm}$, $AC = 5\,\text{cm}$ and $BC = 5\,\text{cm}$.

b Name the type of triangle you have drawn. **E**

2 a Draw an accurate copy of these triangles. **D**

b Write down the length AB for each of the triangles drawn in part **a**.

i Triangle with $CA = 8\,\text{cm}$, angle $C = 45°$, $CB = 11\,\text{cm}$.

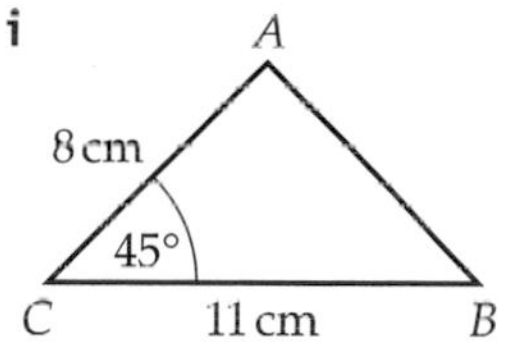

ii Triangle with angle $C = 75°$, angle $B = 40°$, $CB = 9\,\text{cm}$.

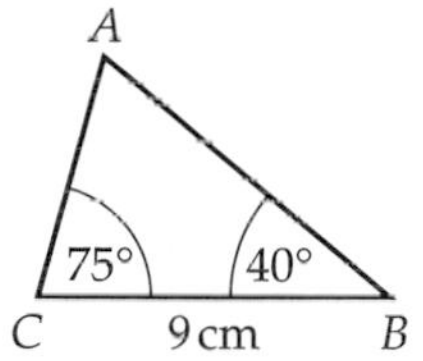

3 a Draw an accurate copy of this triangle. **D**

b Measure and write down the size of $\angle BAC$.

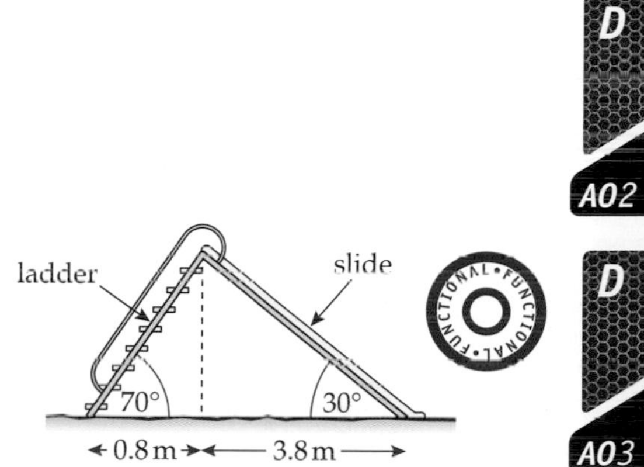

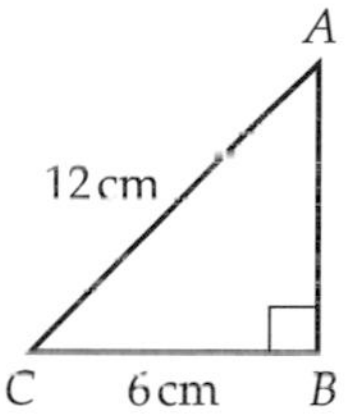

4 An architect is finishing the design of a children's slide. **D** **AO2**
This is the sketch she has made.
The architect says that the length of the ladder is about 2.3 m.
Is the architect correct?

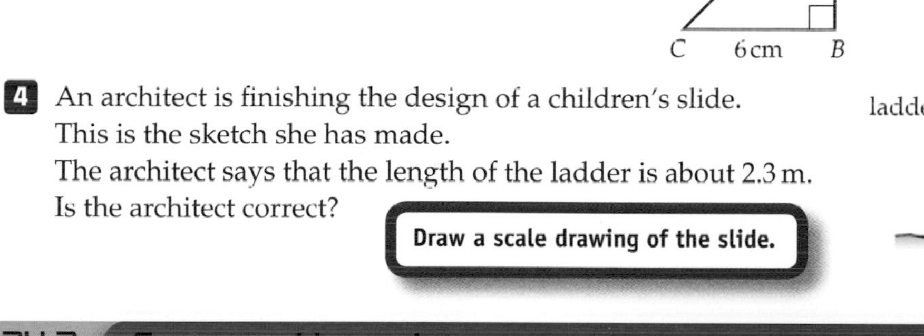

Draw a scale drawing of the slide.

AO3

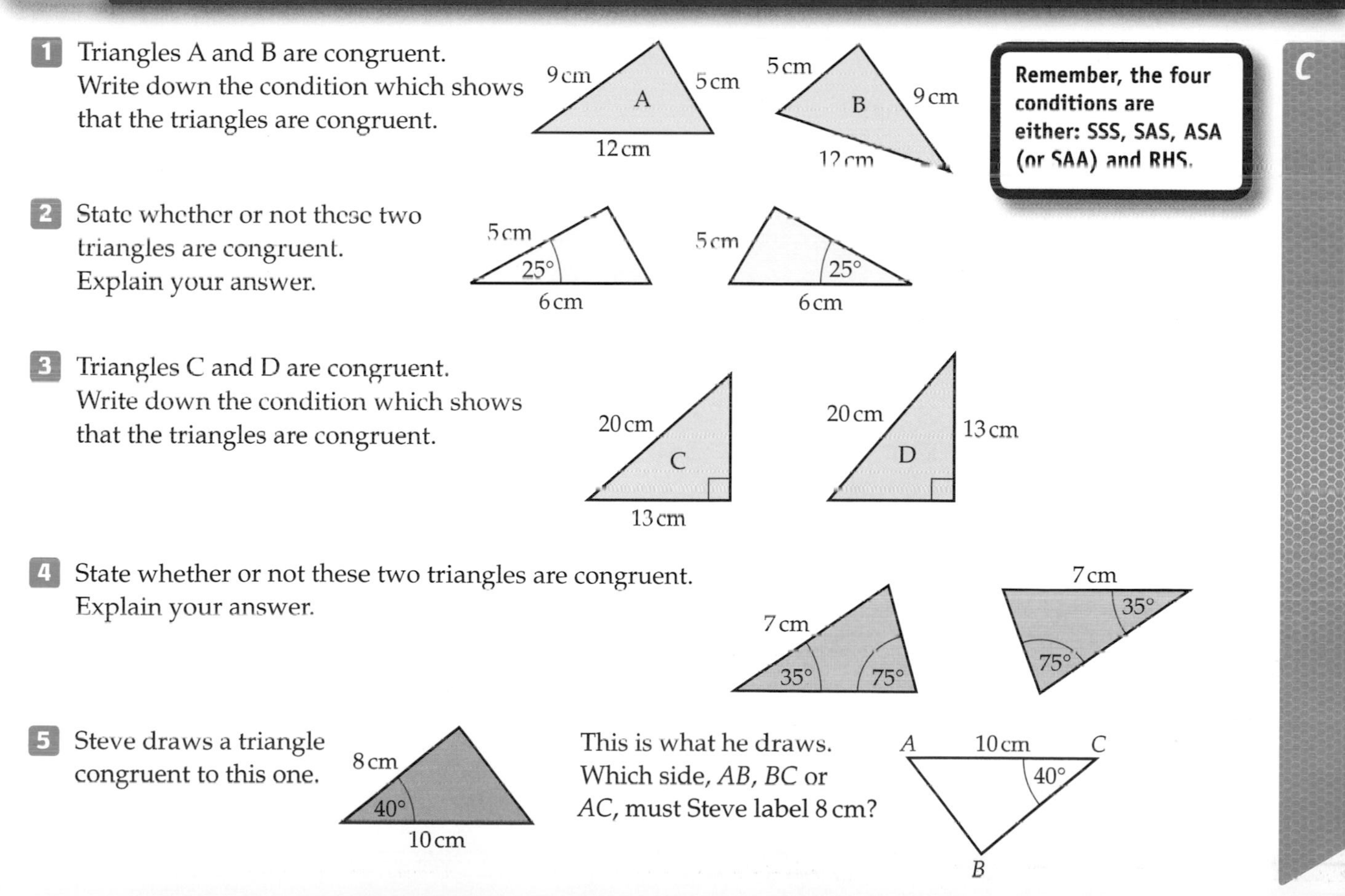

24.3 Congruent triangles

1 Triangles A and B are congruent. **C**
Write down the condition which shows that the triangles are congruent.

Triangle A: 9 cm, 5 cm, 12 cm. Triangle B: 5 cm, 9 cm, 12 cm.

Remember, the four conditions are either: SSS, SAS, ASA (or SAA) and RHS.

2 State whether or not these two triangles are congruent.
Explain your answer.

Triangle 1: 5 cm, 25°, 6 cm. Triangle 2: 5 cm, 25°, 6 cm.

3 Triangles C and D are congruent.
Write down the condition which shows that the triangles are congruent.

Triangle C: 20 cm, 13 cm, right angle. Triangle D: 20 cm, 13 cm, right angle.

4 State whether or not these two triangles are congruent.
Explain your answer.

Triangle 1: 7 cm, 35°, 75°. Triangle 2: 7 cm, 35°, 75°.

5 Steve draws a triangle congruent to this one.

Triangle: 8 cm, 40°, 10 cm.

This is what he draws.
Which side, AB, BC or AC, must Steve label 8 cm?

Triangle: A, C, B; $AC = 10\,\text{cm}$; 40° at C.

Equations, formulae and proof

Links to:
Foundation Student Book
Ch25, pp. 421–430

Key Points

Writing formulae to solve problems D C

You can write word problems using expressions and equations.

You can use letters or words to write your own formulae.

You can substitute values into formulae to solve problems.

Rearranging formulae C

The subject of a formula is the letter on its own.

You can use the rules of algebra to change the subject of the formula.

Changing the subject of a formula is like solving an equation. You need to get a letter on its own on one side of the formula.

Proof C

A proof is a mathematical argument.

When you prove something you need to explain each step of your working.

25.1 Equations and formulae

D

1 Ping-pong balls can be bought in two box sizes. A medium box contains 8 ping-pong balls.
A large box contains 20 ping-pong balls. Jan buys x medium boxes and five large boxes.
In total she buys 132 ping-pong balls.

a Write an equation showing this information.

b Solve your equation to work out the value of x.

2 Leroy thinks of a number. He multiplies his number by 4 then he adds 5. The result is 29.

a Write an equation showing this information.

b Solve your equation to work out the number that Leroy thought of.

C

3 Work out the value of x in this triangle.

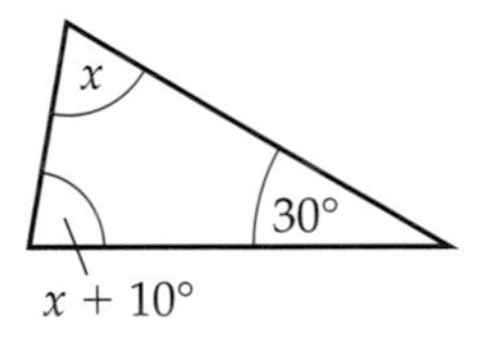

4 The perimeter of this square and regular hexagon are the same.
The length of one side of the hexagon is 8 cm.
Work out the length of one side of the square.

AO2

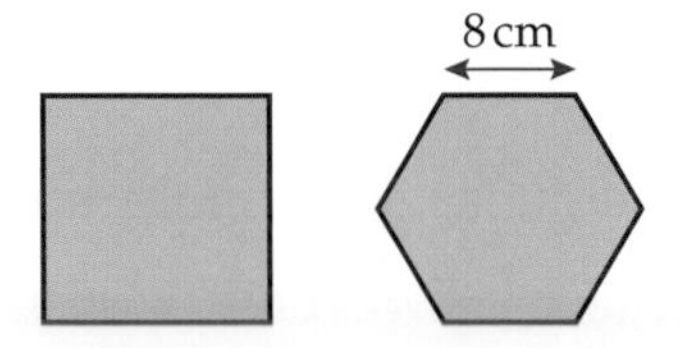

D

5 The length of a rectangle is 5 cm more than its width, w.

a Write a formula for the area of the rectangle, A.

b Work out A when $w = 10$ cm.

C

6 Pens cost p pence each and highlighters cost h pence each.
The total cost of 3 pens and 5 highlighters is C pence.

a Write a formula for C in terms of p and h.

b Calculate p when $C = 250$ and $h = 35$.

7 Joan has n gel pens.
She keeps her 5 favourites, then shares the rest equally among f friends.
Each friend gets G gel pens.

a Write a formula for G in terms of n and f.

b Calculate n when $G = 4$ and $f = 3$.

8 Write down the letter that is the subject of each formula.

a $p = 2 + x$ b $p = 2 - x$ c $7x + 1 = Q$

9 Rearrange the formula $n = 25 - 3x$ to make x the subject.

10 In a sale, rings cost £3 and necklaces cost £4.
Soraya buys x rings and y necklaces. The total cost is £P.

a Write a formula for P in terms of x and y.

b Rearrange your formula to make y the subject.

c Soraya bought 4 rings and spent a total of £20.
How many necklaces did she buy?

11 The formula for the volume of a pyramid is
volume $= \frac{1}{3} \times$ base area $\times$ height.

a Rearrange this formula to make the height the subject.

b Use your rearranged formula to find the height, x, of this pyramid.

Volume $= 48\,\text{cm}^3$, base area $= 36\,\text{cm}^2$

12 The formula for the area of a trapezium is
area $= \frac{1}{2}(a + b)h$.

a Rearrange this formula to make h the subject.

b Use your rearranged formula to find the height, h, of this trapezium.

8 cm, 12 cm, Area $= 45\,\text{cm}^2$

25.2 Proof

1 The diagram shows a rhombus.

a Give a reason why angle a is the same size as angle c.

b Prove that the opposite angles in a rhombus are the same size.

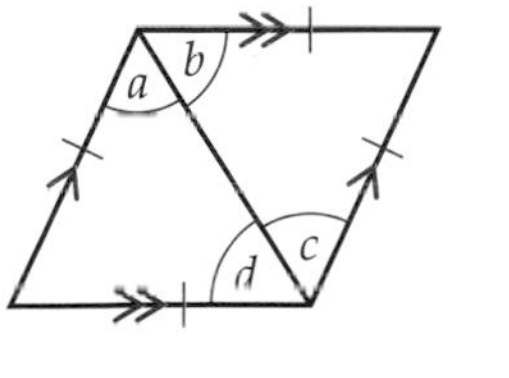

2 Prove that the sum of the interior angles in a hexagon is 720°.

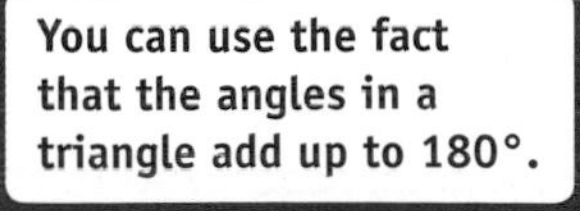

3 Prove that the interior angle in a regular hexagon is 120°.

26 Quadrilaterals and other polygons

Links to:
Foundation Student Book
Ch26, pp. 431–448

Key Points

Rotational symmetry (F)

The order of rotational symmetry is the number of times the shape looks the same in one complete revolution.

Plotting geometric shapes (E)

The coordinates (x, y) give the position of a point on a grid.

The first number in the bracket tells you the position on the x-axis and the second tells you the position on the y-axis.

Lines of symmetry (G F)

A line of symmetry divides a shape into two halves. One half is the mirror image of the other.

Quadrilateral and polygon properties (E D C)

A quadrilateral is a flat shape bounded by four straight lines.

The angle sum of a quadrilateral is 360°.

The sum of the exterior angles of any polygon is 360°.

Exterior angle of a regular polygon $= \dfrac{360°}{\text{number of sides}}$

Number of sides of a regular polygon $= \dfrac{360°}{\text{exterior angle}}$

The sum of the interior angles of any polygon $= (\text{number of sides} - 2) \times 180°$.

Interior angle of a regular polygon $= 180° - \dfrac{360°}{\text{number of sides}}$

Number of sides of a regular polygon $= \dfrac{360°}{(180° - \text{interior angle})}$

26.1 Quadrilaterals and algebra

E

1 Calculate the value of the angles marked with letters.

a
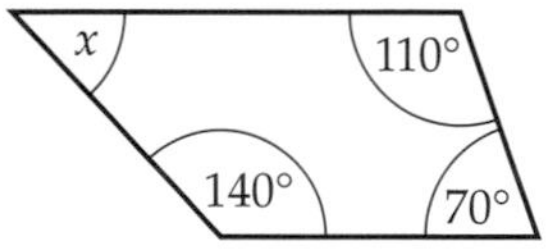

b
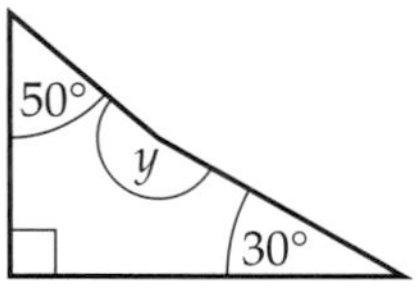

c
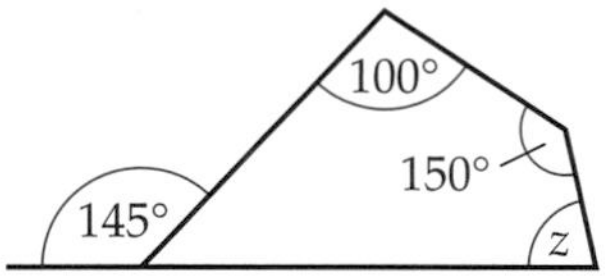

E AO2

2 In this quadrilateral, angle $a = 60°$.
Another angle is twice the size of angle a.
Another angle is half the size of angle a.
What is the size of the remaining angle?

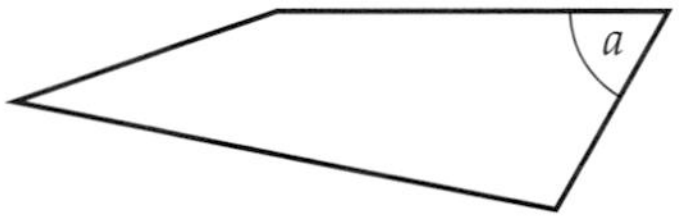

D

3 For each quadrilateral:

a form an equation in x. **b** solve the equation to find the value of x.

i
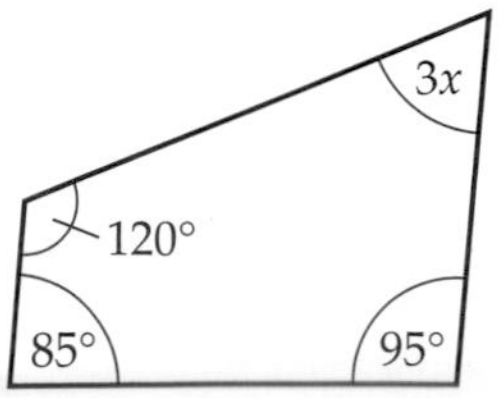

ii
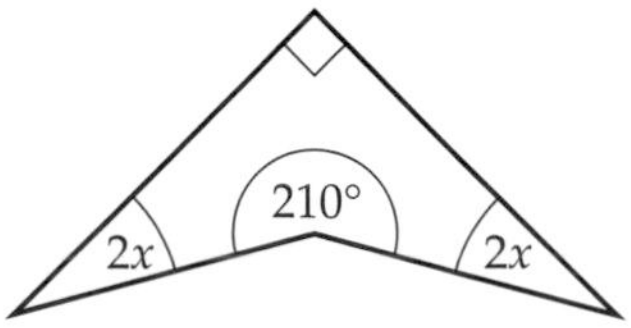

4 A quadrilateral has angles of x, $2x$, $3x$ and $4x$.

a Form an equation in x. **b** Solve the equation to find the value of x.

D AO3

5 Work out the value of $2x + 3y$.

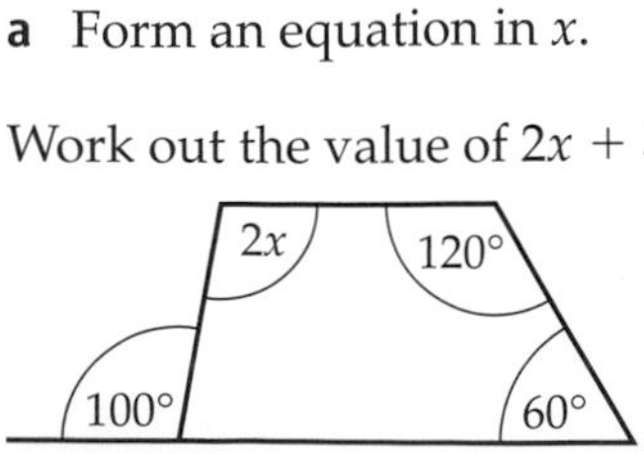

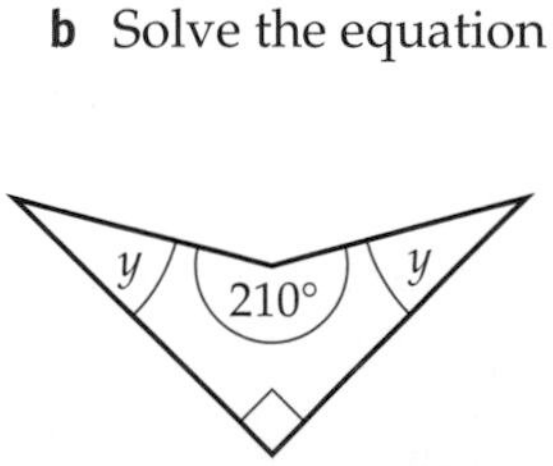

26.2 Special quadrilaterals

1 I am a four-sided shape with no parallel sides. Two of my angles are equal and I've got two pairs of equal sides. My diagonals cut each other at 90°. What shape am I?

F

2 Show how two of these isosceles triangles can be joined together to make a parallelogram.

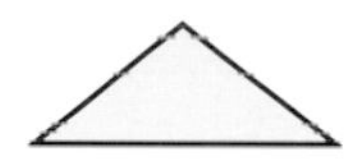

3 Show how you can join together two right-angled triangles to make a kite.

4 Calculate the value of the angles marked with letters.

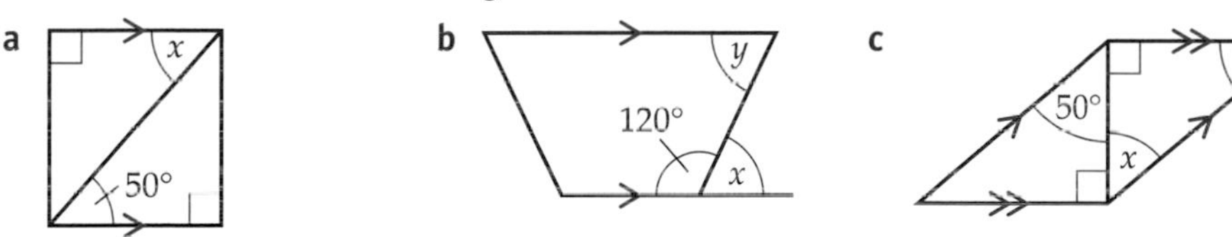

5 Calculate the value of the angles marked with letters.

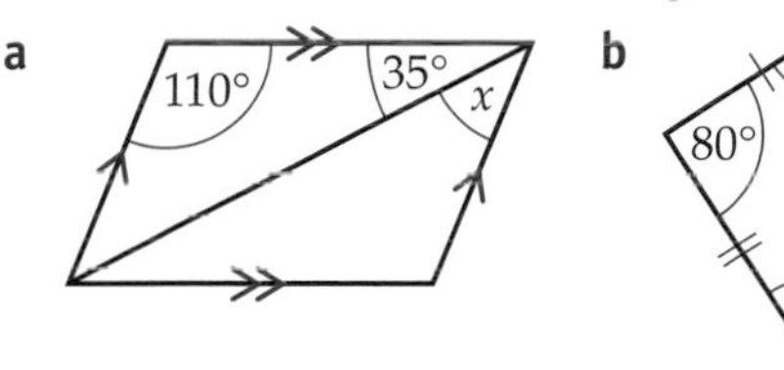

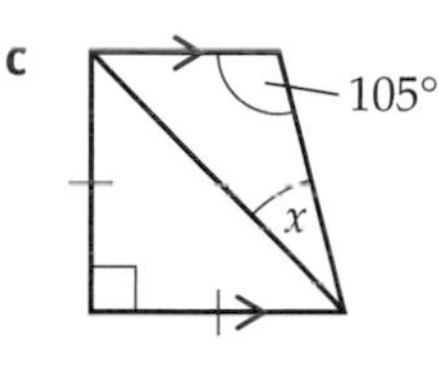

c 105° x

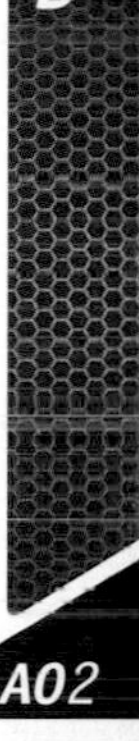

6 Jack says, 'In this chevron angle x is 130°.'
Marta says, 'You're wrong. Angle x is 100°.'
Who is correct?
Show working to support your answer.

130° x

A02

26.3 Polygons 1

1 Calculate the size of angle x in each of these polygons.

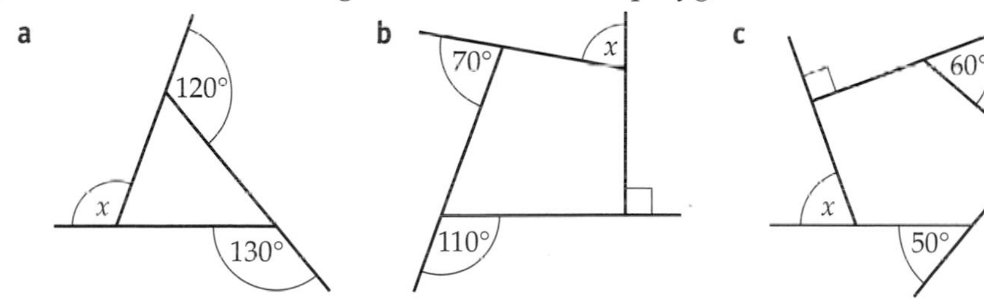

2 The exterior angles of this polygon are a, b, c and d.
Angle b is 10° more than angle a.
Angle c is twice the size of angle a.
Angle d is three times the size of angle a.
How many degrees is angle a?

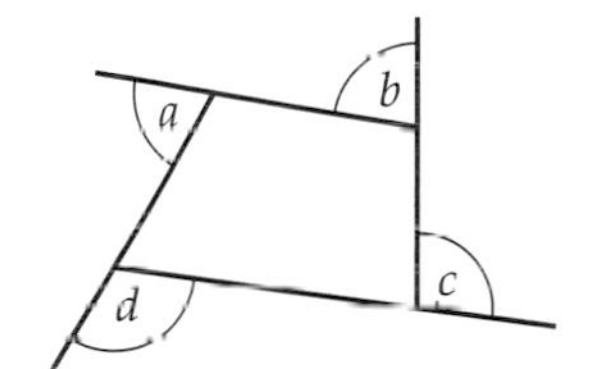

3 Calculate the size of the exterior angle of a regular polygon with 12 sides.

C

4 Explain why it is not possible for the exterior angle of a regular polygon to be 42°.

D

1 Calculate the size of angle x in each of these polygons.

a

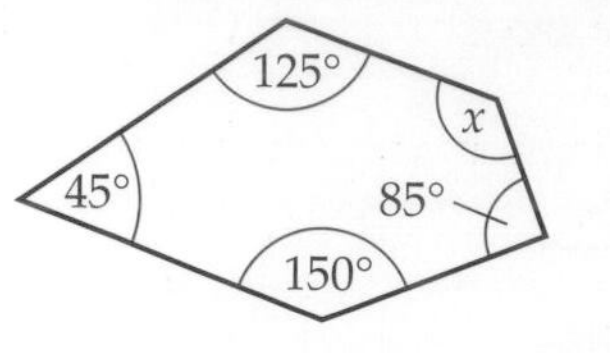

b

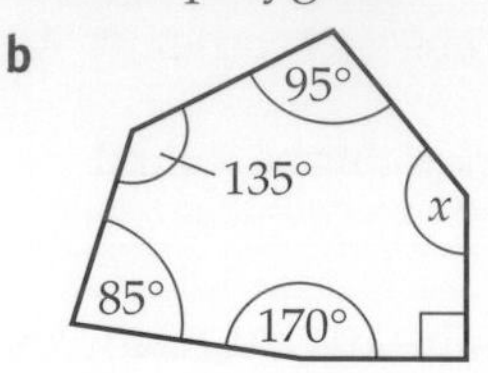

2 Calculate the value of y in this polygon.

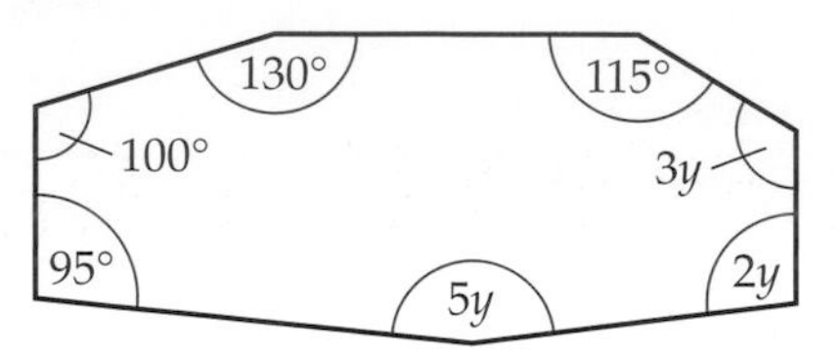

C

3 Calculate the size of the interior angle of a regular polygon with 12 sides.

4 How many sides does a regular polygon have if the interior angle is 160°?

5 Work out the sizes of angles x, y and z in this regular octagon.

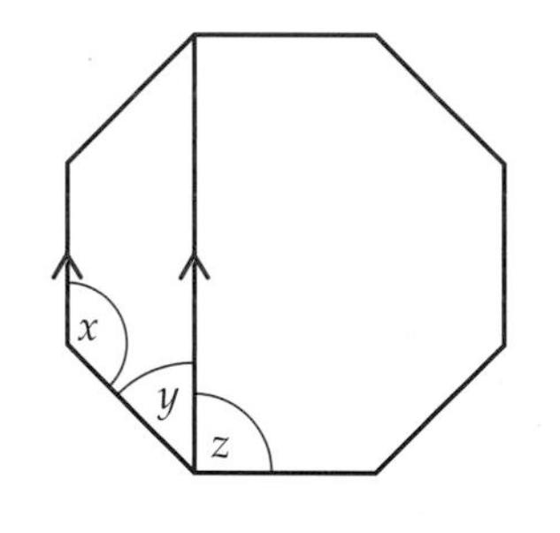

C

A02

6 A carpenter is making a wooden seat to go around a tree.
The seat is in the shape of a regular pentagon.
The diagram shows a plan view of the seat.
At what angle, marked as x on the diagram, must the carpenter cut the wood for the seat?

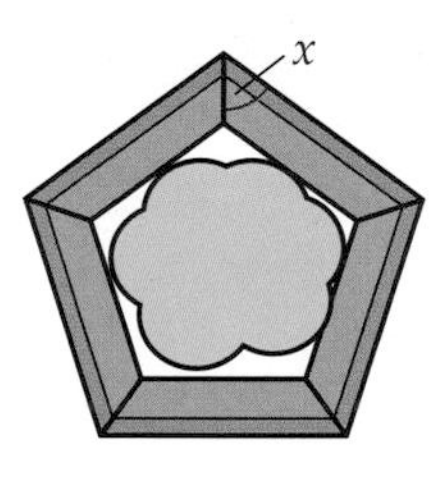

26.5 Coordinates

E

1 **a** Write down the coordinates of A.

b Write down the coordinates of B.

c Write down the coordinates of C.

Copy the diagram.

d Mark the midpoint of AC with a cross and label it E.

e Write down the coordinates of E.

f Plot the point $D(-1, -1)$ on your diagram.

g Join A to B to C to D to A.

h What is the mathematical name of the quadrilateral $ABCD$?

y
4
B
3
2
C
1
A
−4 −3 −2 −1 O 1 2 3 4 x
−1
−2
−3
−4

2 Sanjay plots the three points shown on the grid.
Sanjay says, 'If I plot another point at (0, 3) I'll get a kite.'
Essien says, 'No, you won't, you'll get a square.'
Who is correct? Explain your answer.

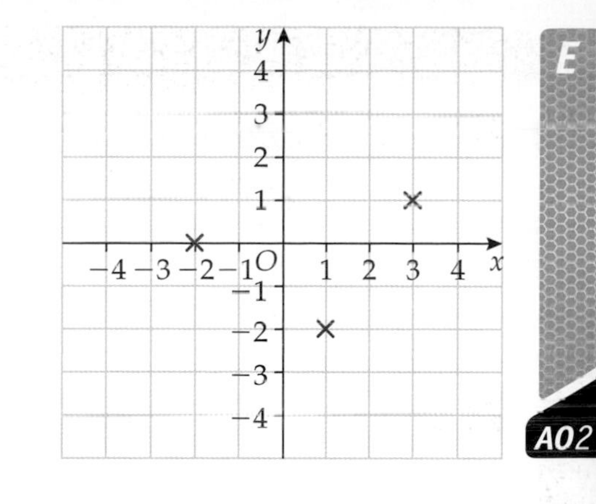

26.6 Symmetry

1 Copy each of these shapes and draw on any lines of symmetry.

a 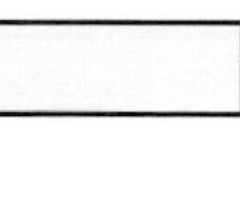
b
c 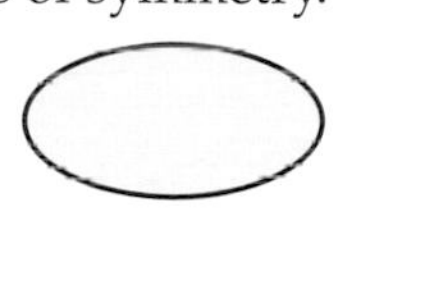
d 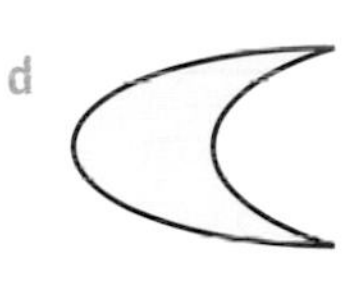

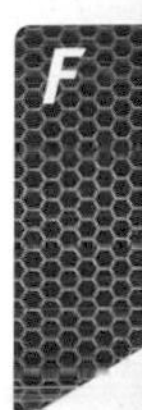

2 Write down the number of lines of symmetry that each of these shapes has.

a 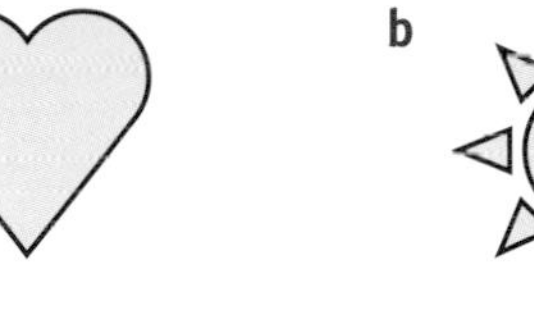
b
c 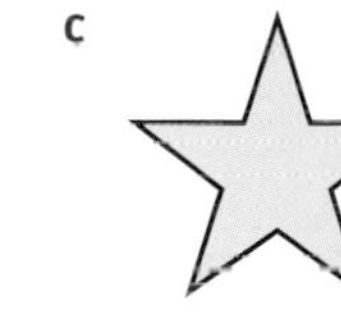
d

3 Copy this grid.
Shade in three more squares so that the final pattern has two lines of symmetry.

4 Write down the order of rotational symmetry of the shapes in Q1.

5 For each of these shapes write down:

i the number of lines of symmetry

ii the order of rotational symmetry.

a 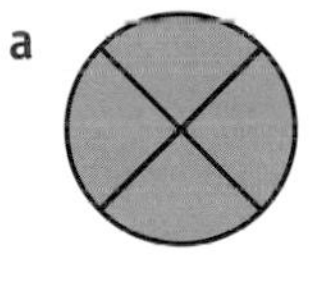
b 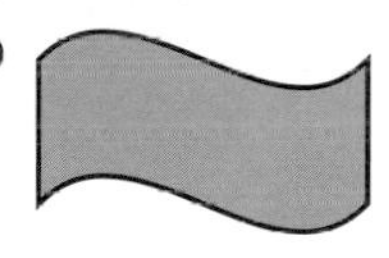
c 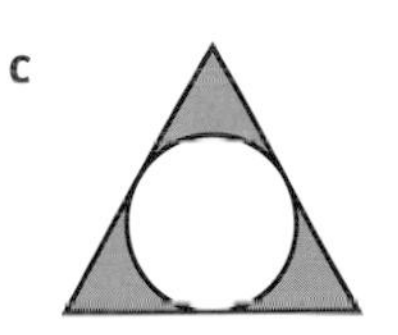

6 What is the smallest number of squares that need to be shaded to give this shape rotational symmetry of order 2?
Explain your answer.

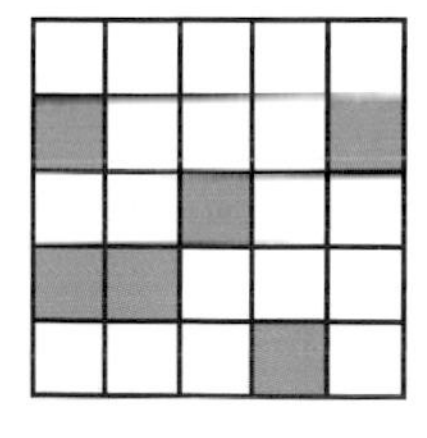

Links to:
Foundation Student Book
Ch27, pp. 449–456

Key Points

Conversions between metric and imperial units

E

You need to remember these for your exam.

Metric	8 km	2.5 cm	4.5 litres	1 kg
Imperial	5 miles	1 inch	1 gallon	2.2 pounds

Scale drawings and maps

E

The scale tells you the relationship between lengths on the drawing or map and lengths in real life.

Scales for maps are usually given as ratios.

For example, a ratio of 1 : 25 000 means that 1 cm on the map represents 25 000 cm in real life.

27.1 Metric and imperial units

E

1 Copy and complete.

a 5 miles ≈ _______ km

b 10 miles ≈ _______ km

c _______ miles ≈ 32 km

2 Copy and complete.

a 1 inch ≈ _______ cm

b 5 inches ≈ _______ cm

c _______ inches ≈ 20 cm

3 Copy and complete.

a 1 kg ≈ _______ pounds

b 6 kg ≈ _______ pounds

c _______ kg ≈ 22 pounds

4 Copy and complete.

a 1 gallon ≈ _______ litres

b 3 gallons ≈ _______ litres

c _______ gallons ≈ 36 litres

5 Sebastian is driving in France.
He sees this sign.
How many miles is Sebastian from Paris?

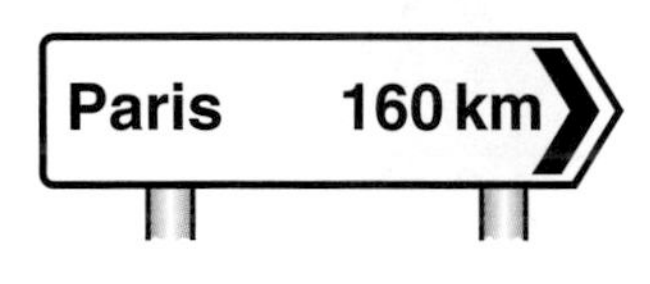

6 Harry sees an advertisement for a chain saw.
The chain saw has a 14-inch blade.
How many centimetres is this?

E

7 Nadia is making bread. Her recipe uses $1\frac{1}{2}$ kg of flour.
Nadia only has 2½ pounds of flour.
How much more flour does Nadia need?

8 A lift can carry a maximum of 350 kg.
Three people weighing 16 stone, 18 stone and 17½ stone are waiting to get into the lift.
There are 14 pounds in 1 stone.

A02

a What is the total weight of the people in pounds?

b Should all three of them get in the lift? Explain your answer.

E

1 A scale drawing uses a scale of 1 cm to represent 25 m.
Work out the length on the drawing of each of these real-life lengths.

a 50 m b 225 m

2 A scale drawing uses a scale of 1 cm to represent 50 m.
Work out the length in real-life of each of these lengths on the drawing.

a 3 cm b 6.5 cm

3 Jana has made a scale drawing of her office on cm-squared paper.

a How wide is the window?

b What are the length and width of the storage cupboard?

c Jana has a fridge that is 50 cm wide.
Can she fit the fridge between the storage cupboard and the workstation?

d Jana wants to put a conference table in the centre of her office.
She has a table that measures 2.4 m long and 1.2 m wide.
She wants a space at least 60 cm wide all the way around the table.
Will her table fit in the office?
Explain your answer.

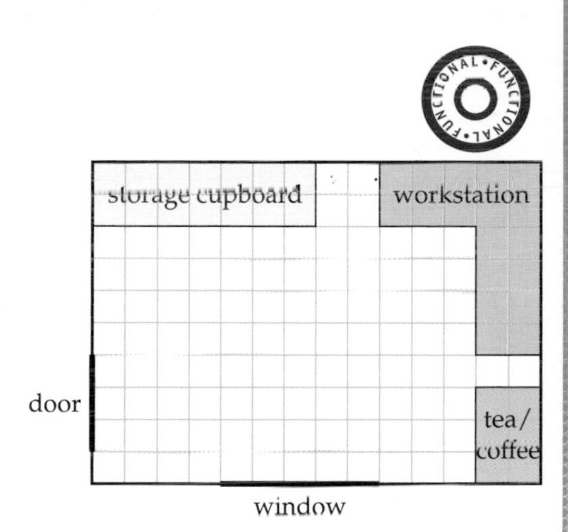

Scale: 1 cm represents 30 cm.

4 Andy competing in a sponsored cycle ride.
The diagram shows a map of the cycle route.

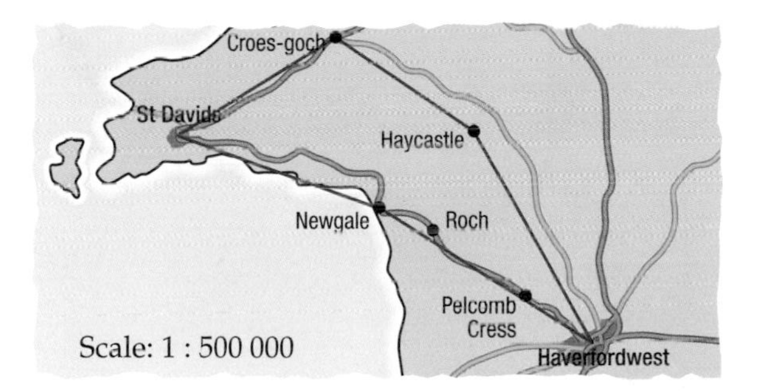

a Convert 500 000 cm into km.

b What distance would be represented by 5 cm on the map?

c What length on the map would represent a distance of 1 km?

d Measure the distance from Croes-goch to St Davids.

e Work out the real-life distance from Croes-goch to St Davids.

f The route of the cycle ride is shown in red on the map.
Work out an approximate length of the whole cycle ride.

Links to:
Foundation Student Book
Ch28, pp. 460–471

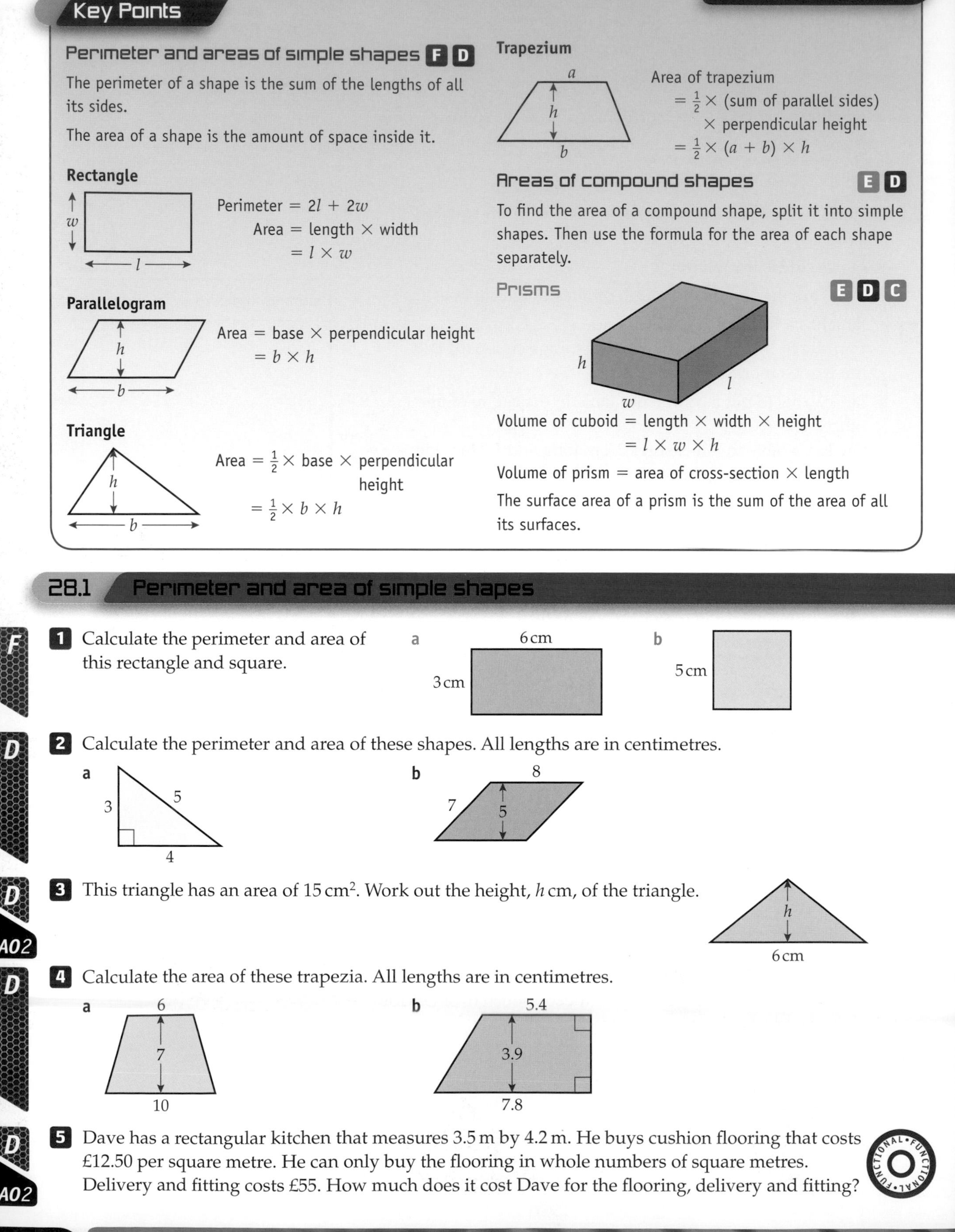

Key Points

Perimeter and areas of simple shapes F D

The perimeter of a shape is the sum of the lengths of all its sides.

The area of a shape is the amount of space inside it.

Rectangle

Perimeter $= 2l + 2w$

Area = length × width $= l \times w$

Parallelogram

Area = base × perpendicular height $= b \times h$

Triangle

Area $= \frac{1}{2} \times$ base × perpendicular height $= \frac{1}{2} \times b \times h$

Trapezium

Area of trapezium $= \frac{1}{2} \times$ (sum of parallel sides) × perpendicular height $= \frac{1}{2} \times (a + b) \times h$

Areas of compound shapes E D

To find the area of a compound shape, split it into simple shapes. Then use the formula for the area of each shape separately.

Prisms E D C

Volume of cuboid = length × width × height $= l \times w \times h$

Volume of prism = area of cross-section × length

The surface area of a prism is the sum of the area of all its surfaces.

28.1 Perimeter and area of simple shapes

F

1 Calculate the perimeter and area of this rectangle and square.

a 6 cm, 3 cm **b** 5 cm

D

2 Calculate the perimeter and area of these shapes. All lengths are in centimetres.

a 3, 4, 5 **b** 8, 7, 5

D **AO2**

3 This triangle has an area of 15 cm^2. Work out the height, h cm, of the triangle.

h, 6 cm

D

4 Calculate the area of these trapezia. All lengths are in centimetres.

a 6, 7, 10 **b** 5.4, 3.9, 7.8

D **AO2**

5 Dave has a rectangular kitchen that measures 3.5 m by 4.2 m. He buys cushion flooring that costs £12.50 per square metre. He can only buy the flooring in whole numbers of square metres. Delivery and fitting costs £55. How much does it cost Dave for the flooring, delivery and fitting?

FUNCTIONAL

28.2 Perimeter and area of compound shapes

1 a Calculate the perimeter of this compound shape.

b Calculate the area of this compound shape by first dividing the shape into two rectangles, A and B.

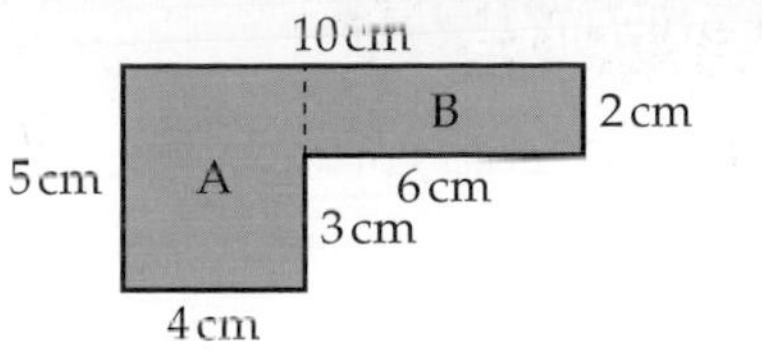

2 Calculate the perimeter and area of each of these compound shapes. All lengths are in centimetres.

a

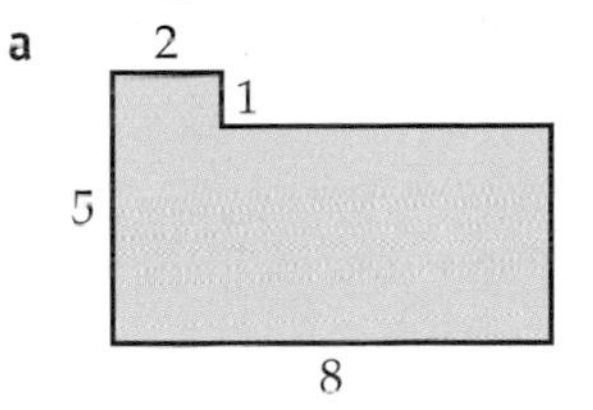

b

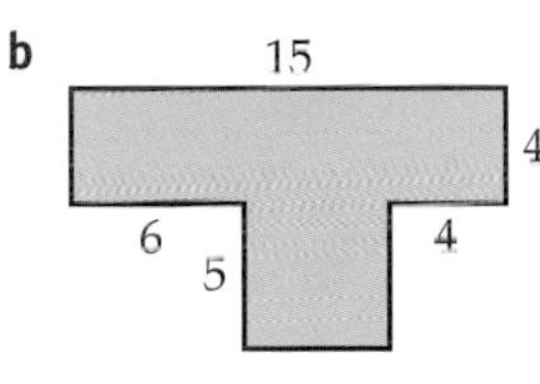

3 Calculate the perimeter and area of each of these compound shapes. All lengths are in centimetres.

a

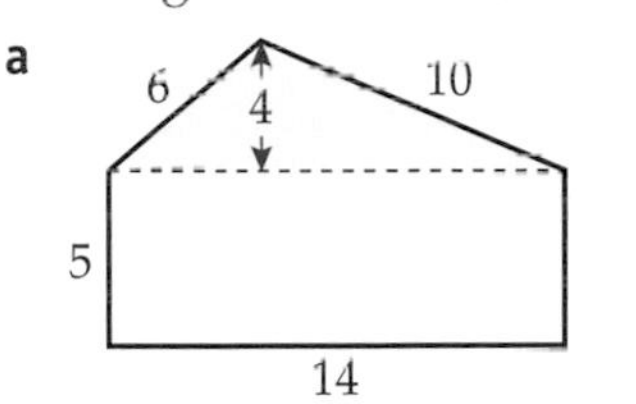

b

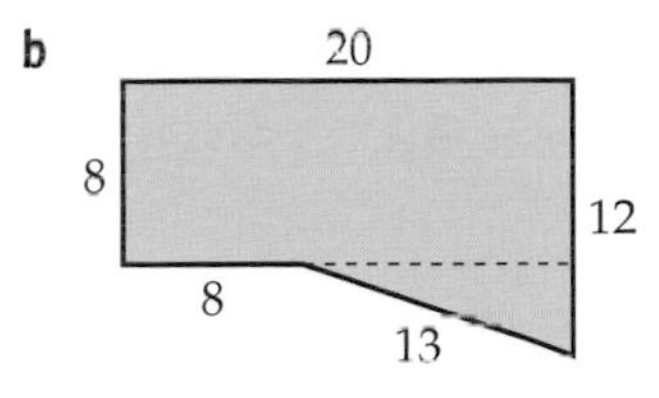

28.3 Volume and surface area of prisms

1 Calculate the volume of this cube and cuboid.

a

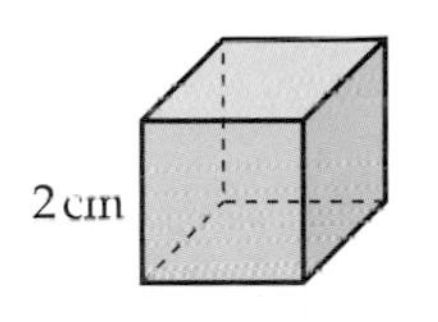

b

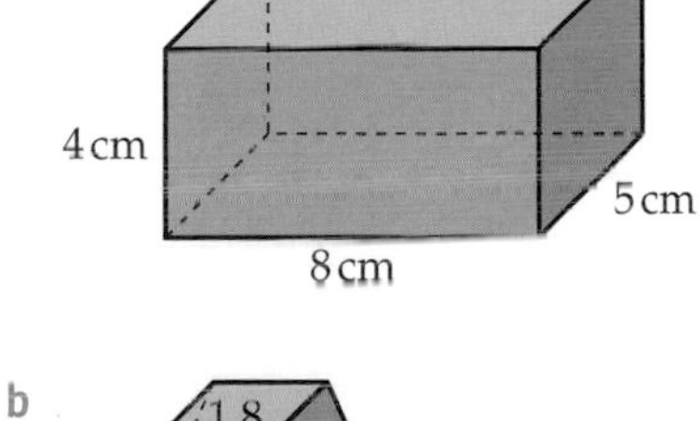

2 Calculate the volume of these prisms.

a

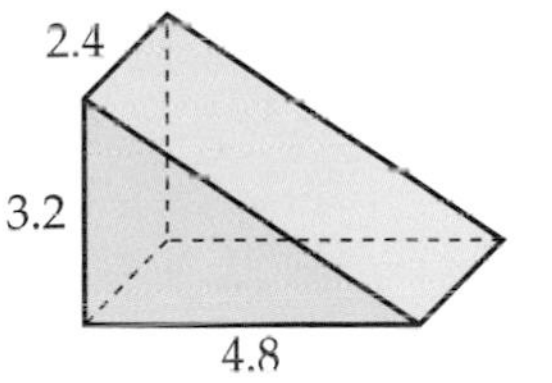

b

1.8
3
2
4.5

3 Jamie is the manager of a swimming pool.
The dimensions of the swimming pool are shown on the diagram. To clean the pool Jamie adds 100 g of chlorine for every 50 m^3 of water.
How many kg of chlorine must Jamie add to the pool?

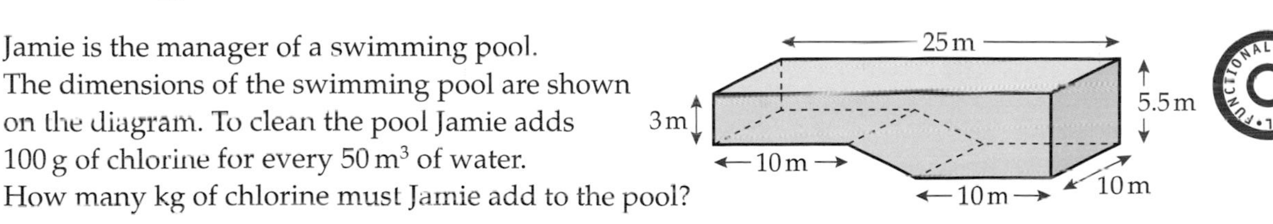

4 Calculate the surface area of these prisms.

a

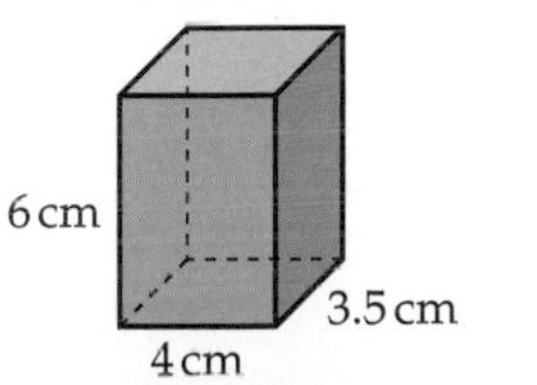

b

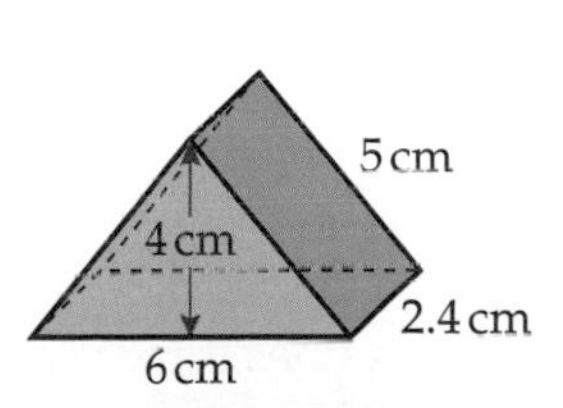

Links to:
Foundation Student Book
Ch29, pp. 472–480

Key Points

Net of a 3-D object G F

A net is a 2-D representation of a 3-D object. The net will fold up to make the 3-D object.

Isometric drawing E

Draw along the printed lines of the paper.

Vertical lines on the paper represent the vertical lines of the object.

The lines at an angle on the paper represent the horizontal lines on the object.

Plane of symmetry D

When a 3-D object is cut along a plane of symmetry, it will be cut into two identical halves.

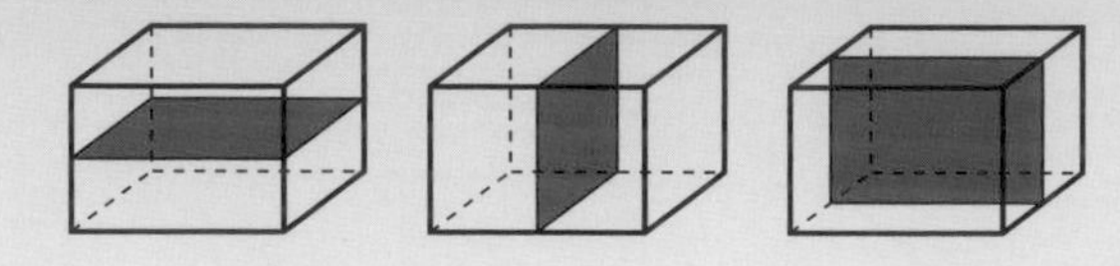

Plans and elevations D

The plan is the view from above the object.

The front elevation is the view from the front of the object.

The side elevation is the view from the side of the object.

29.1 Nets of 3-D objects

G

1 Which of the following could be the net of a cube?

F

2 Here is the net of a 3-D object.
Sketch the object.

3 Make accurate drawings of the nets of these 3-D objects.

4 Make accurate drawings of two possible nets of this 3-D object.

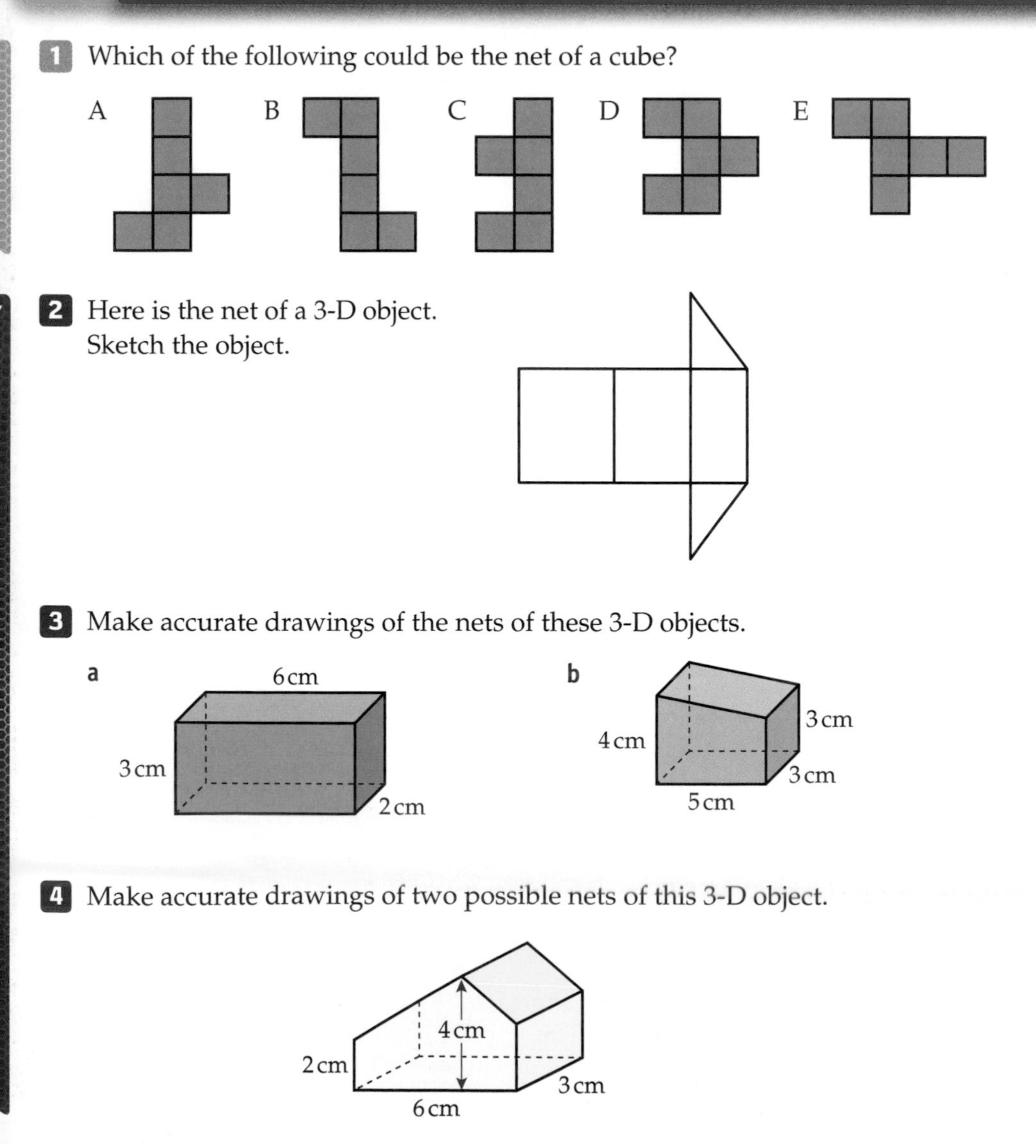

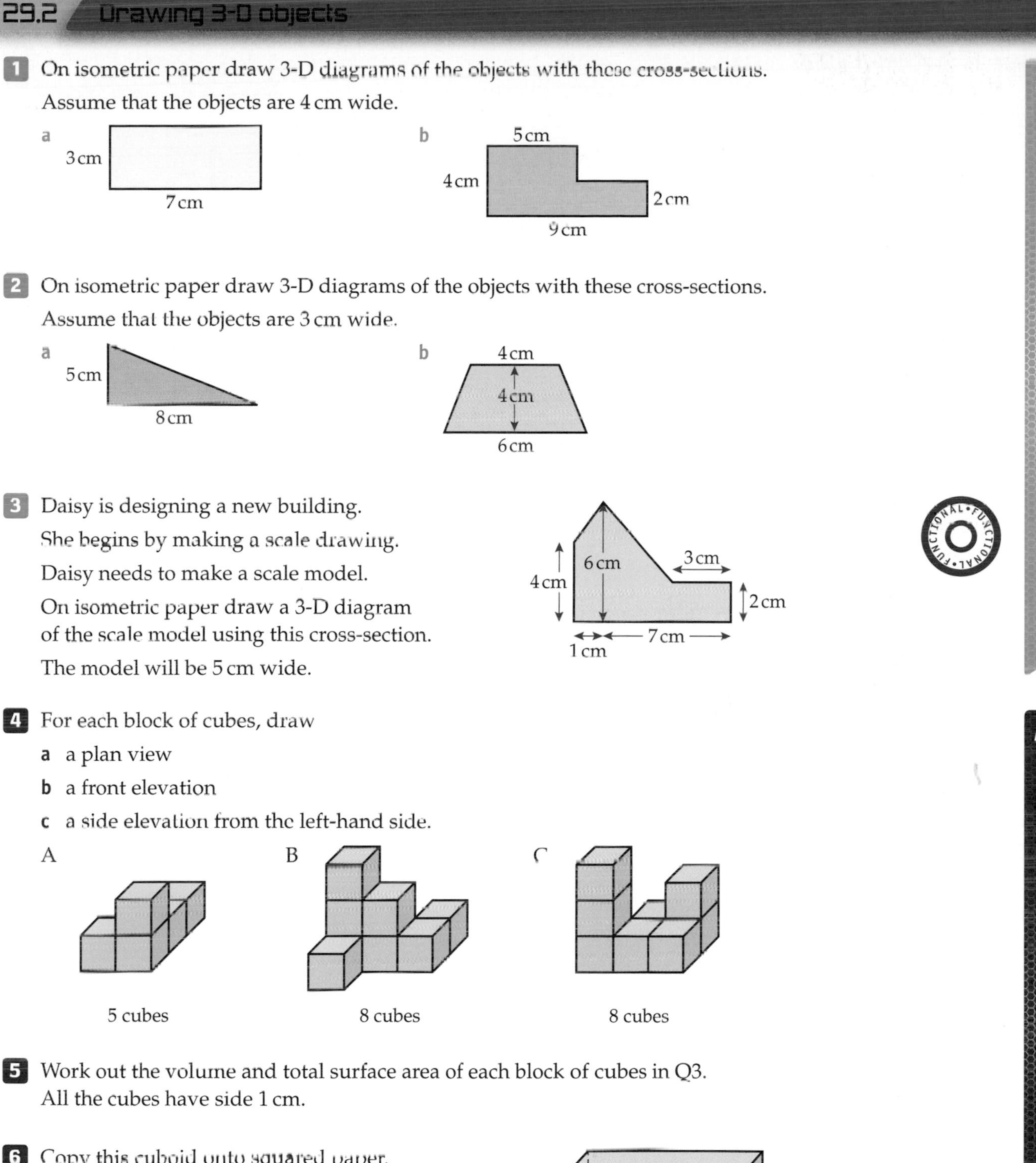

1 On isometric paper draw 3-D diagrams of the objects with these cross-sections.
Assume that the objects are 4 cm wide.

a

b

2 On isometric paper draw 3-D diagrams of the objects with these cross-sections.
Assume that the objects are 3 cm wide.

a

b

3 Daisy is designing a new building.
She begins by making a scale drawing.
Daisy needs to make a scale model.
On isometric paper draw a 3-D diagram of the scale model using this cross-section.
The model will be 5 cm wide.

4 For each block of cubes, draw

a a plan view

b a front elevation

c a side elevation from the left-hand side.

5 Work out the volume and total surface area of each block of cubes in Q3.
All the cubes have side 1 cm.

6 Copy this cuboid onto squared paper.
Show all its planes of symmetry.
Draw separate diagrams for each plane of symmetry.

7 Copy this object onto squared paper.
Show all its planes of symmetry.
Draw separate diagrams for each plane of symmetry.

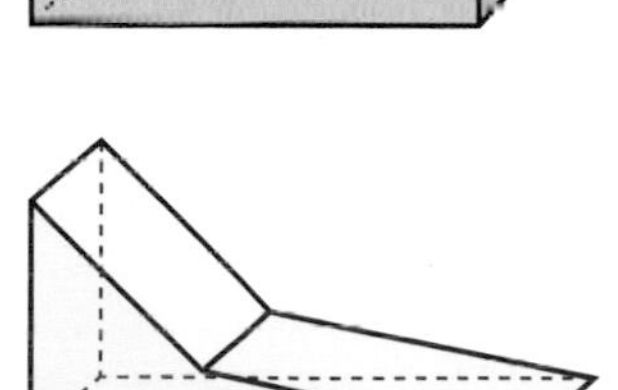

Reflection, translation and rotation

Links to:
Foundation Student Book Ch30, pp. 481–497

Key Points

Reflection
G F E D C

To describe a reflection on a grid, you need to give the equation of the mirror line.

Rotation

To describe a rotation fully, you need to give

- the centre of the rotation
- the size of turn
- the direction of turn.

Translation
D C

To describe a translation, you have to give the distance and direction of movement.

You can describe a translation using a column vector:

$\begin{pmatrix}3\\2\end{pmatrix}$ means move 3 in the x-direction and then move 2 in the y-direction.

Combined transformations
C

Reflection, translation and rotation are transformations. They transform an object to an image. For these transformations, the object and its image are congruent.

30.1 Reflection

G

1 For each part

i copy each diagram onto squared paper

ii draw the reflection of each shape in the given mirror line.

a

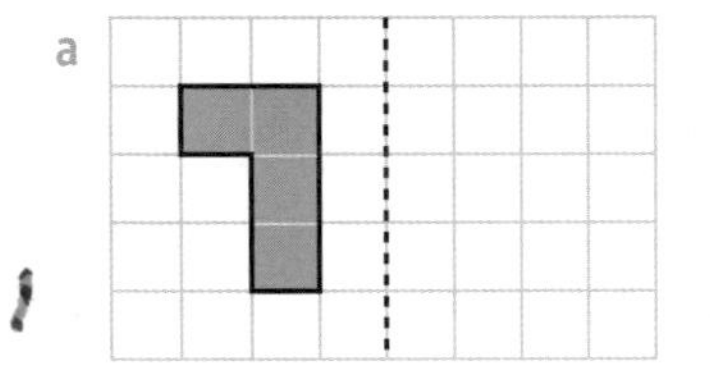

b

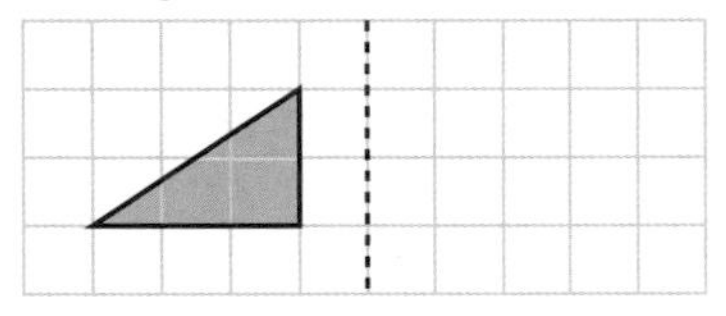

c

F

2 For each part

i copy the diagram onto squared paper

ii draw the reflection of each shape in the given mirror line.

a

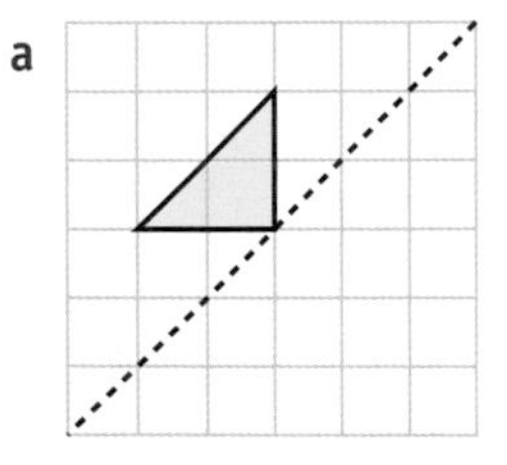

b

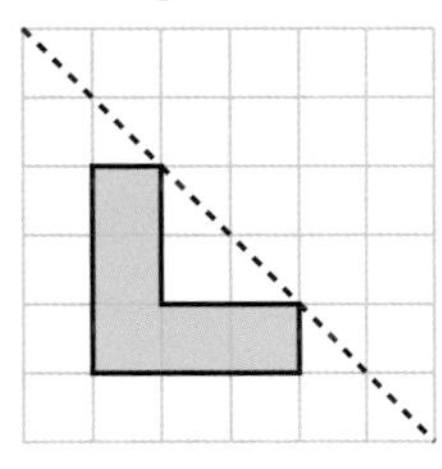

c

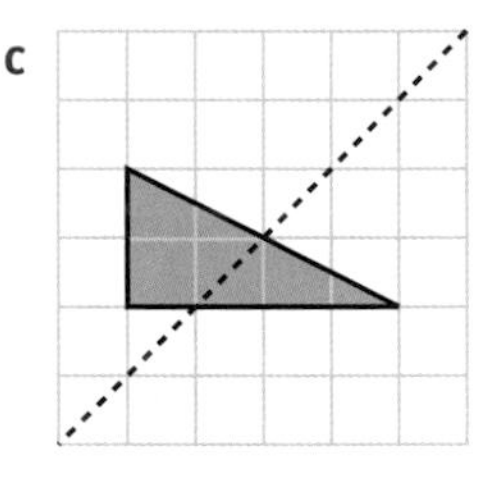

E

3 Copy this coordinate grid.

a Draw the reflection of triangle A in the y-axis.
Label the reflected shape C.

D

b Draw the reflection of triangle A in the line $y = 3$.
Label the reflected shape D.

c Draw the reflection of triangle B in the line $x = -1$.
Label the reflected shape E.

d Complete this sentence.
Triangle B is a reflection of triangle A in the line ________.

A02

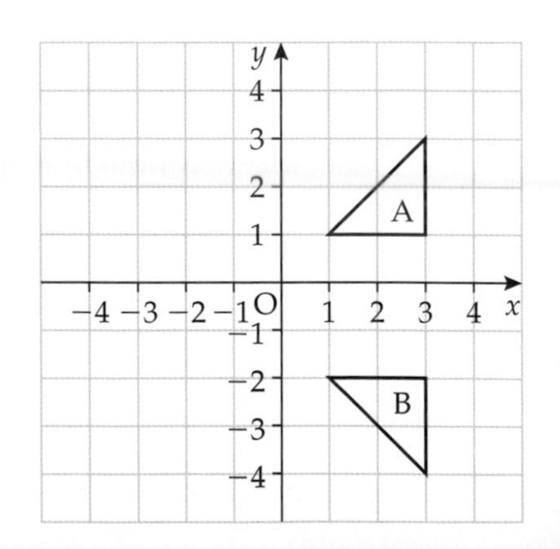

4 Describe the transformation that takes

a shape A to shape B

b shape A to shape C.

C

AO2

30.2 Translation

1 Copy the following shapes onto squared paper.
Draw the image of the shape after the given translation.

D

a

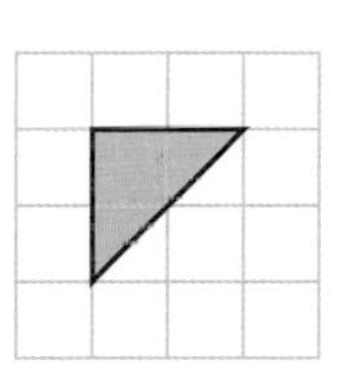

2 squares right
1 square down

b

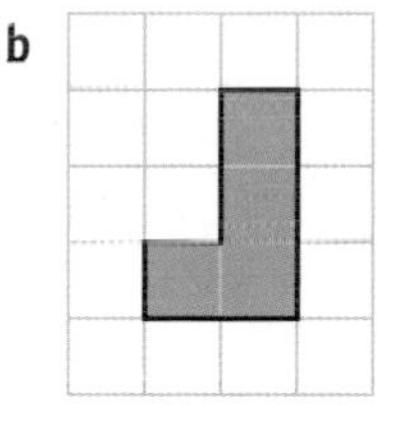

1 square left
3 squares up

c

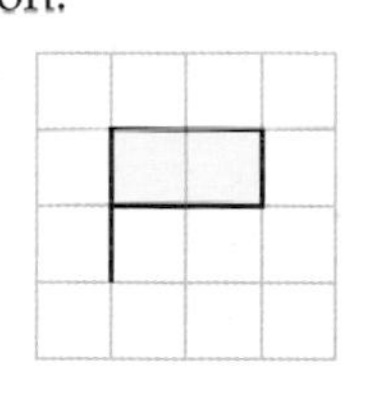

3 squares left
2 squares down

2 On this grid, shape A is translated to shape B.
Describe the translation.

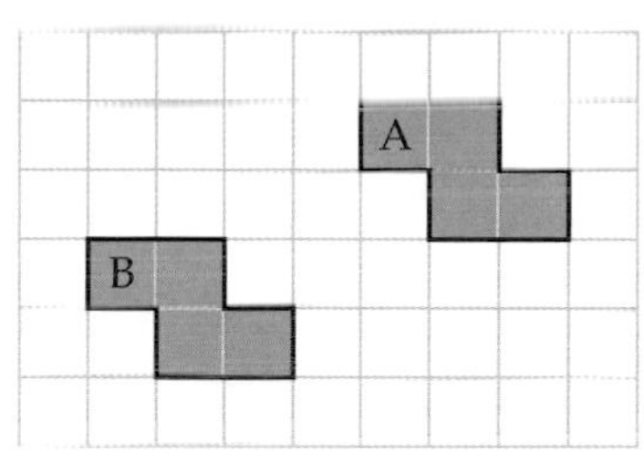

D

AO2

3 Write down the column vector for each of the following translations shown on this coordinate grid.

a Triangle A to triangle B

b Triangle B to triangle C

c Triangle C to triangle A

d Triangle A to triangle C

e What do you notice about your answers in parts **c** and **d**?

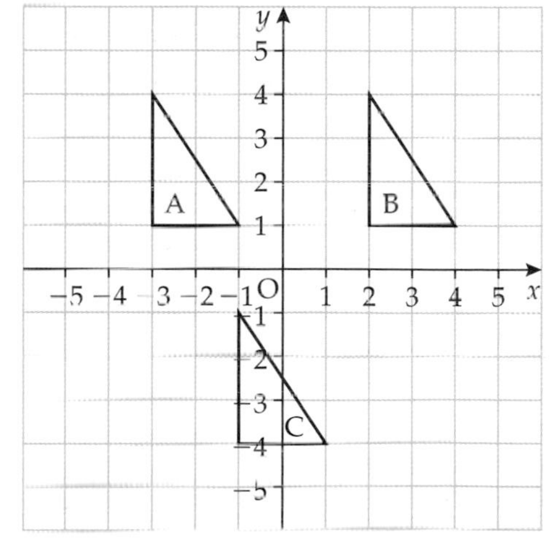

C

AO2

4 On a grid, shape A is translated to shape B by the vector $\begin{pmatrix} 2 \\ 3 \end{pmatrix}$ and shape B is translated to shape C by the vector $\begin{pmatrix} 3 \\ -5 \end{pmatrix}$.

a Write down the column vector that translates shape A directly to shape C.

b Write down the column vector that translates shape C directly to shape A.

C

AO3

D

1 Copy each shape and the centre of rotation on to squared paper.
Draw the image of the shape after the rotation given.

a

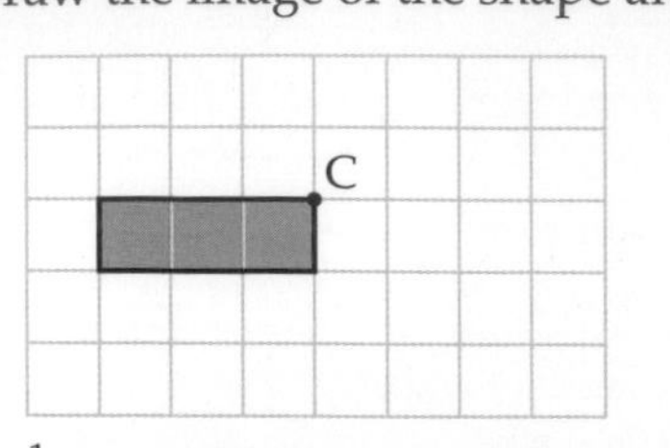

$\frac{1}{4}$ turn anticlockwise about centre C

b

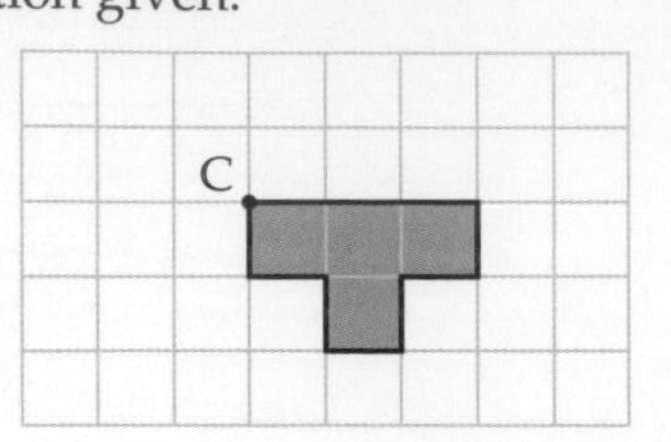

$\frac{1}{2}$ turn clockwise about centre C

For Q2–4, copy this coordinate grid and draw shapes A, B and C.

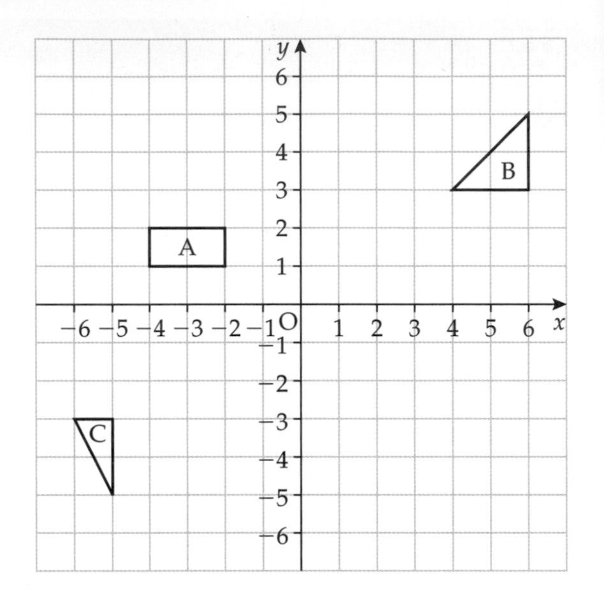

2 Draw the image of shape A after a rotation of 90° clockwise about the point (0,0). Label the image A′.

C

3 Draw the image of shape B after a rotation of 180° about the point (3,1). Label the image B′.

C AO3

4 Rotate shape C about the point (−4, −3) to make a pattern with rotational symmetry of order 4.

D

For Q5 and Q6 use this coordinate grid.

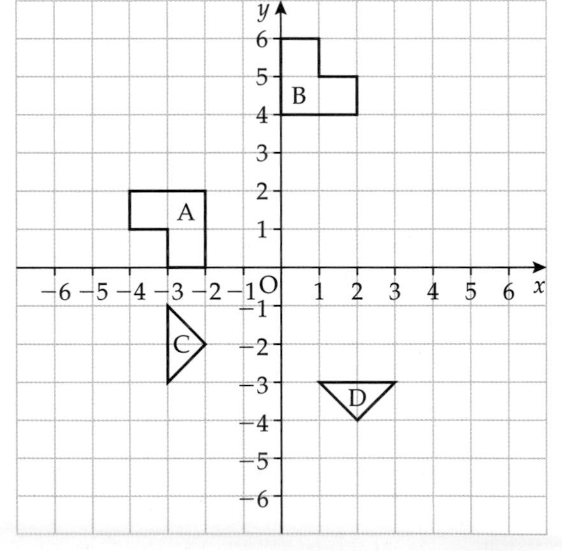

5 Shape A rotates on to shape B.

a What size turn is the rotation?

b What direction is the rotation?

c Find the centre of rotation and write down the coordinates.

Use tracing paper to help.

C AO2

6 Describe fully the transformation that maps shape C on to shape D.

C AO3

7 A children's fairground wheel has eight seats.
The seats are at the end of arms which are spaced equally around a centre hub.
Each seat and arm looks like this.

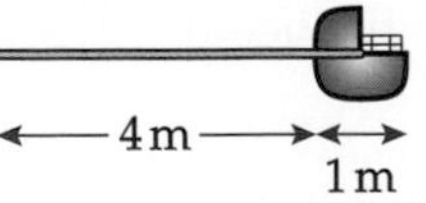

Make a scale drawing of the wheel on a coordinate grid.
Use a scale of 1 grid square to 1 metre.

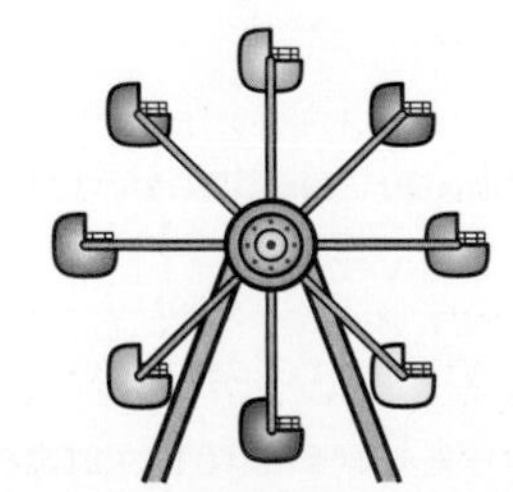

31 Circles and cylinders

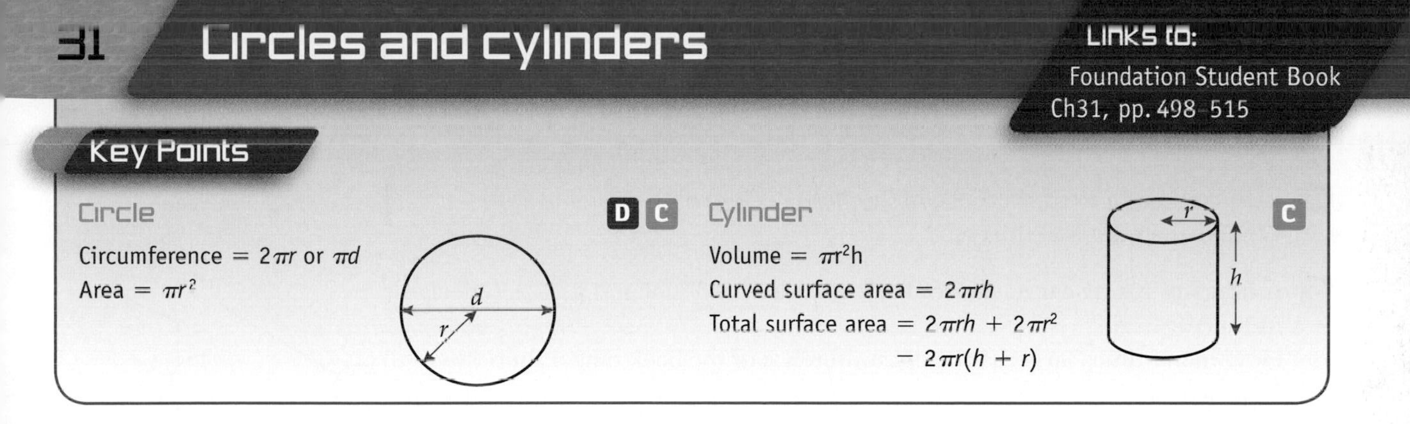

31.1 Basic circle work

G

1 Bobby has a circular piece of paper. He folds the paper in half and opens it out again.
The line through the centre is a mirror line.
Give another mathematical name for this line.

2 a Draw a circle of radius 4 cm.
b On any point on the circumference of the circle draw a circle of diameter 4 cm.
c Draw a line between the two points where the circumferences cross.
i What is the name of this straight line?
ii What is the name of the curved line that is between the two points where the circles cross?
d Draw a line between the two points where the circumferences cross, to the centre of one of the circles. What is the name of this line?
e Draw one line that is a tangent to both circles.

31.2 Circumference of a circle

For these questions, use the π button on your calculator unless you are asked to leave your answer in terms of π.

D

1 Beverly builds a circular scalextric track.
a The inside track has a radius of 120 cm.
Find, to 1 d.p., the distance around this track.
b The outside track has a diameter of 260 cm.
Find, to 1 d.p., the distance around this track.
c How much further is the outside track than the inside track?
d Calculate the circumference of a track of radius 2.1 m.
Leave your answer in terms of π.

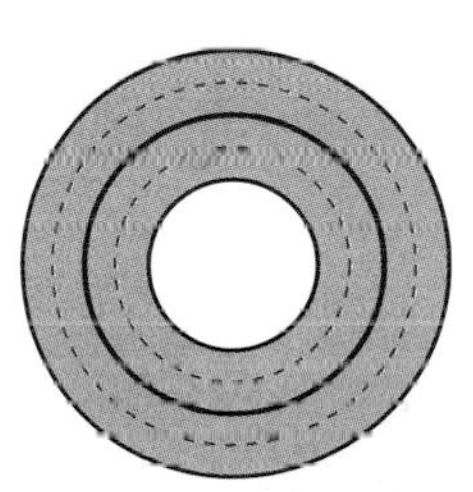

2 A hamster wheel has a diameter of 16 cm.
How many complete revolutions does the wheel make for the hamster to run a distance of 10 m?

3 A 10p coin has a circumference of 7.5 cm.
Find the diameter of a 10p coin.
Give your answer to 3 s.f.

4 A 1p coin has a circumference of 2π cm.
What is the radius of the coin?

5 A dragster racing car has a front wheel radius of 35 cm and a back wheel radius of 85 cm.
How many times larger is the circumference of the back wheel than the front wheel.

6 Calculate the perimeter of this paving slab.
Give your answer to the nearest cm.

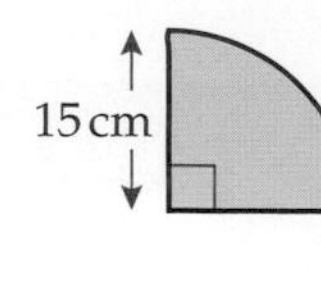

Remember to include the straight edges.

7 A circular pattern made from paving slabs is shown.
The thick grey lines are where cement has to be placed to lock the slabs together.
The cement is called grout.
Work out the total length of grout needed for this pattern of slabs.
Give your answer to the nearest cm.

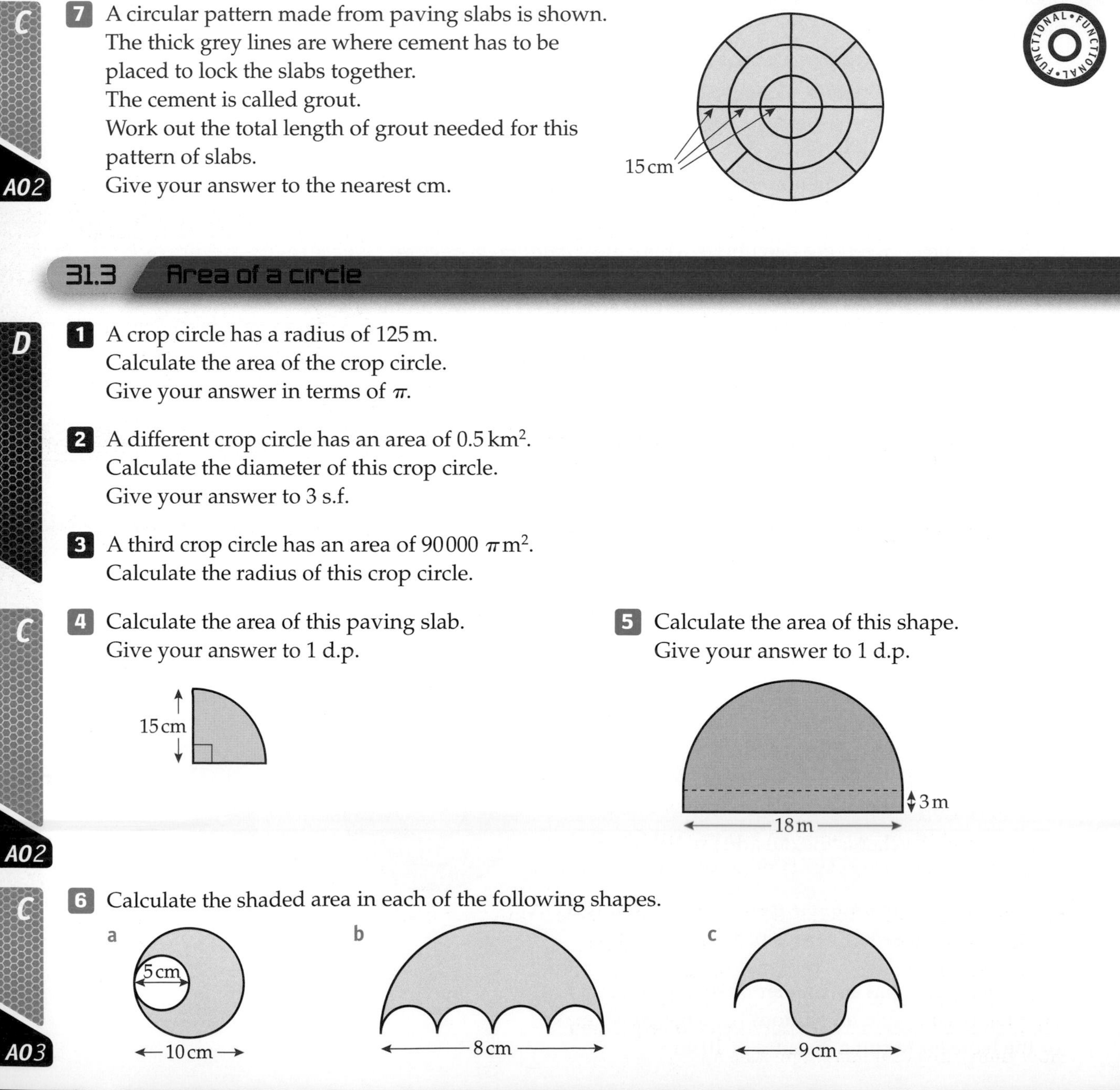

31.3 Area of a circle

1 A crop circle has a radius of 125 m.
Calculate the area of the crop circle.
Give your answer in terms of π.

2 A different crop circle has an area of 0.5 km^2.
Calculate the diameter of this crop circle.
Give your answer to 3 s.f.

3 A third crop circle has an area of $90\,000\ \pi$ m^2.
Calculate the radius of this crop circle.

4 Calculate the area of this paving slab.
Give your answer to 1 d.p.

5 Calculate the area of this shape.
Give your answer to 1 d.p.

6 Calculate the shaded area in each of the following shapes.

a

b

c

7 The radius of each circle in this shape is 5 cm.
What area of the shape is shaded blue?
Give your answer to 3 s.f.

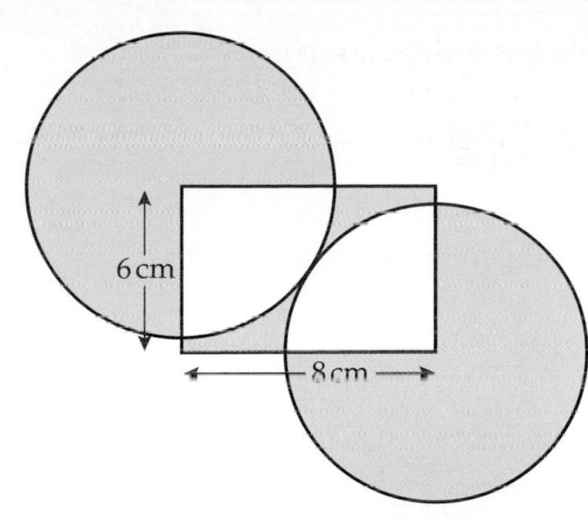

31.4 Cylinders

1 A cylindrical tea urn has a radius of 25 cm and height 60 cm.

a Calculate the volume of the tea urn.
Give your answer in terms of π.

b Use the fact that 1 litre $= 1000\text{ cm}^3$ to work out how many complete litres the tea urn holds.

2 Cylindrical tins of chocolates are packed into a box.
The dimensions of a tin and a box are shown in the diagram.

a What is the most number of tins that can be packed into a box?

b What is the volume of one tin?
Give your answer in terms of π.

c What is the total volume of the box?

d What is the total volume of all the tins in the box?

e What volume of the box is empty?
Give your answer to 3 s.f.

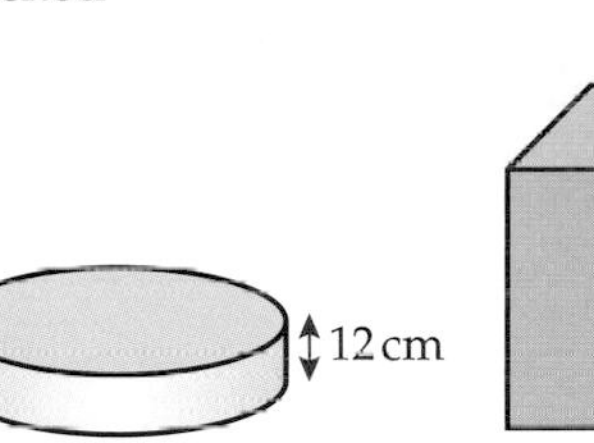

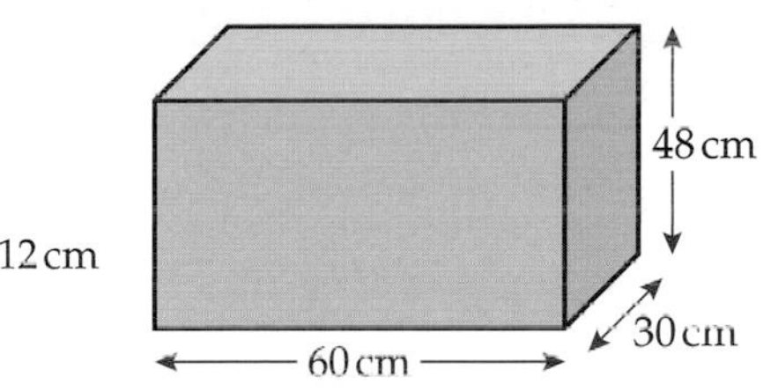

3 Which of these two cylinders has the largest capacity, and by how much?
Give your answer to the nearest m*l*.

4 A cylinder of height 4.5 m has a volume of 905 m^3.
Work out the radius of the cylinder correct to 1 d.p.

5 A factory makes concrete cylinder seats to go outside a shopping mall.
Each cylinder is 36 cm high with a radius of 45 cm.
Each visible face of the cylinder is painted.
A 5-litre pot of paint covers approximately 9 m^2.
Altogether there are 28 cylinders seats to be painted.
How many tins of paint are needed?

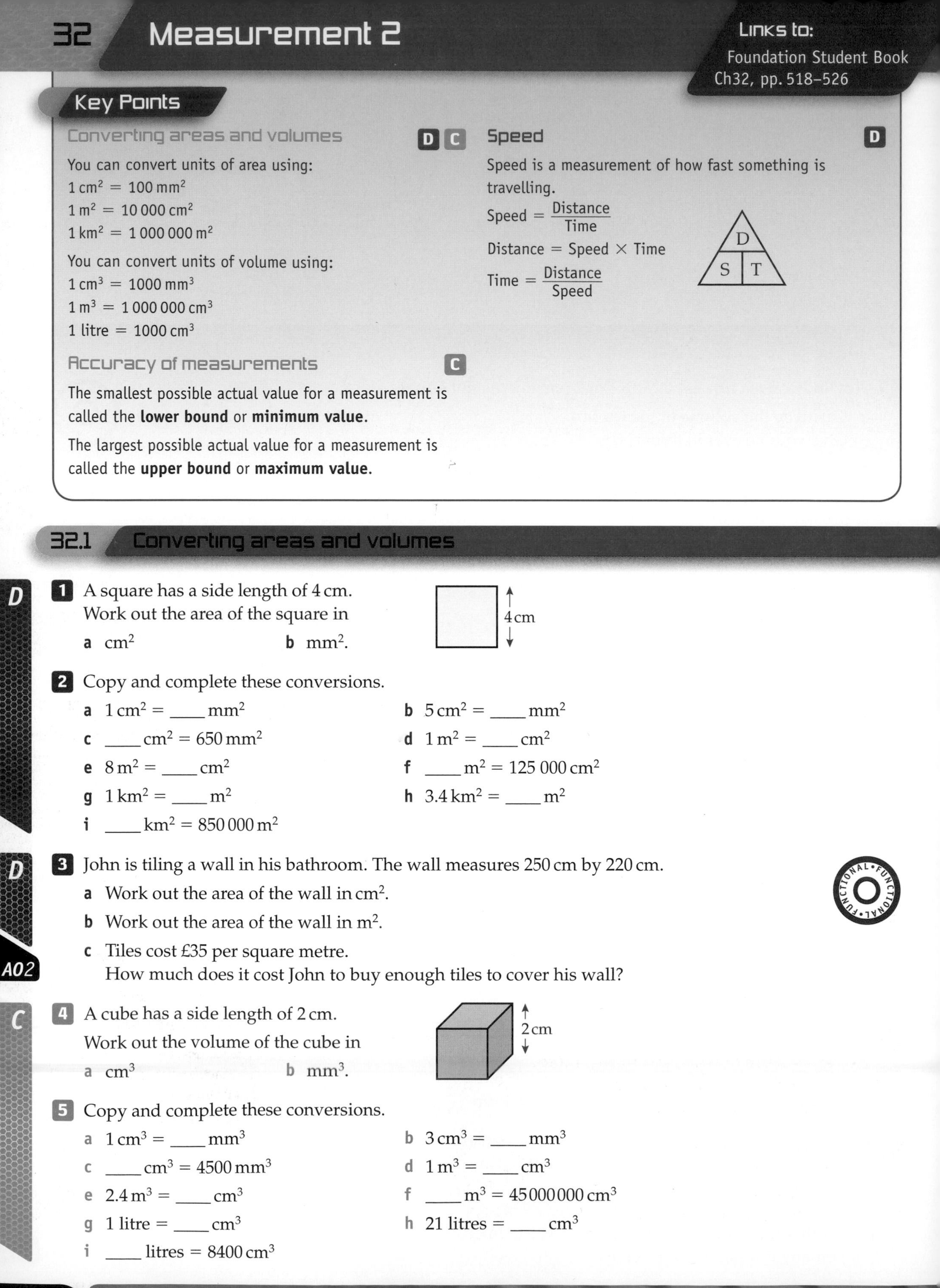

Links to: Foundation Student Book Ch32, pp. 518–526

Key Points

Converting areas and volumes (D C)

You can convert units of area using:

$1\text{ cm}^2 = 100\text{ mm}^2$

$1\text{ m}^2 = 10\,000\text{ cm}^2$

$1\text{ km}^2 = 1\,000\,000\text{ m}^2$

You can convert units of volume using:

$1\text{ cm}^3 = 1000\text{ mm}^3$

$1\text{ m}^3 = 1\,000\,000\text{ cm}^3$

$1\text{ litre} = 1000\text{ cm}^3$

Accuracy of measurements (C)

The smallest possible actual value for a measurement is called the **lower bound** or **minimum value.**

The largest possible actual value for a measurement is called the **upper bound** or **maximum value**.

Speed (D)

Speed is a measurement of how fast something is travelling.

$\text{Speed} = \dfrac{\text{Distance}}{\text{Time}}$

$\text{Distance} = \text{Speed} \times \text{Time}$

$\text{Time} = \dfrac{\text{Distance}}{\text{Speed}}$

32.1 Converting areas and volumes

D

1 A square has a side length of 4 cm.
Work out the area of the square in

a cm^2 **b** mm^2.

2 Copy and complete these conversions.

a $1\text{ cm}^2 = \_\_\_\_\text{ mm}^2$

b $5\text{ cm}^2 = \_\_\_\_\text{ mm}^2$

c $\_\_\_\_\text{ cm}^2 = 650\text{ mm}^2$

d $1\text{ m}^2 = \_\_\_\_\text{ cm}^2$

e $8\text{ m}^2 = \_\_\_\_\text{ cm}^2$

f $\_\_\_\_\text{ m}^2 = 125\,000\text{ cm}^2$

g $1\text{ km}^2 = \_\_\_\_\text{ m}^2$

h $3.4\text{ km}^2 = \_\_\_\_\text{ m}^2$

i $\_\_\_\_\text{ km}^2 = 850\,000\text{ m}^2$

D

AO2

3 John is tiling a wall in his bathroom. The wall measures 250 cm by 220 cm.

a Work out the area of the wall in cm^2.

b Work out the area of the wall in m^2.

c Tiles cost £35 per square metre.
How much does it cost John to buy enough tiles to cover his wall?

C

4 A cube has a side length of 2 cm.
Work out the volume of the cube in

a cm^3 **b** mm^3.

5 Copy and complete these conversions.

a $1\text{ cm}^3 = \_\_\_\_\text{ mm}^3$

b $3\text{ cm}^3 = \_\_\_\_\text{ mm}^3$

c $\_\_\_\_\text{ cm}^3 = 4500\text{ mm}^3$

d $1\text{ m}^3 = \_\_\_\_\text{ cm}^3$

e $2.4\text{ m}^3 = \_\_\_\_\text{ cm}^3$

f $\_\_\_\_\text{ m}^3 = 45\,000\,000\text{ cm}^3$

g $1\text{ litre} = \_\_\_\_\text{ cm}^3$

h $21\text{ litres} = \_\_\_\_\text{ cm}^3$

i $\_\_\_\_\text{ litres} = 8400\text{ cm}^3$

6 Paulo has a water tank in the loft in his house.
The water tank is a cuboid measuring 615 mm by 305 mm by 550 mm.
The water tank is full when Paulo needs to empty it.
He empties the tank of water into 14-litre buckets.
How many buckets does he need?

FUNCTIONAL · C · AO3

32.2 Accuracy of measurements

1 Sinead cycles to work every morning.
The distance she cycles is 4 miles, to the nearest mile.
Write down

a the upper bound for this distance
b the lower bound for this distance.

C

2 A square has a side length of 8 cm, to the nearest cm.
Work out

a the minimum side length of the square
b the minimum area of the square.

C · AO2

3 Fola times how long it takes her to complete a Sudoku puzzle.
She takes 52.7 seconds, correct to one decimal place.
Work out

a the upper bound for the time
b the lower bound for the time.

C

4 The lengths of this triangle are given to one decimal place.
Work out the maximum area of the triangle.

5.8 cm
9.0 cm

C · AO3

32.3 Speed

1 Copy and complete:

a distance = speed × ________
b speed = $\frac{\text{distance}}{}$
c time = $\frac{}{\text{speed}}$

D

2 David drives 240 km in 3 hours.
Work out his average speed.

3 A coach travels at an average speed of 70 km/h on a journey of 245 km.
How long did the journey take?

4 Shen runs at an average speed of 5 m/s.
How far does she run in 12 seconds?

Remember m/s means metres per second.

5 Harry uses a machine to paint the white lines on a football pitch.
He paints 100 m of white line in $3\frac{1}{2}$ minutes.
What is Harry's average speed in metres per second?
Give your answer correct to 2 d.p.

Start by changing $3\frac{1}{2}$ minutes into seconds.

D

6 Lowri is driving in France.
She sees this speed limit sign.
The sign is in km/h.
What is the speed limit in mph?

120

FUNCTIONAL · AO2

Links to:
Foundation Student Book
Ch33, pp. 527–535

Key Points

Enlargement

F E D

An enlargement changes the size of an object but not its shape.

The number of times each length of a shape is enlarged is called the scale factor.

In an enlargement, all the angles stay the same but all the lengths change in the same proportion. The image is similar to the object.

To describe an enlargement fully you must give the scale factor and the centre of enlargement.

33.1 Enlargement

F

1 **a** Shape A is enlarged to get shape B.
What is the scale factor of the enlargement?

b Shape C is enlarged to get shape D.
What is the scale factor of the enlargement?

A B C D

E

2 a By counting squares, write down the perimeter of this shape.

b Copy the shape onto squared paper.

c Enlarge the shape by a scale factor of 4.

d What is the perimeter of the enlargement?

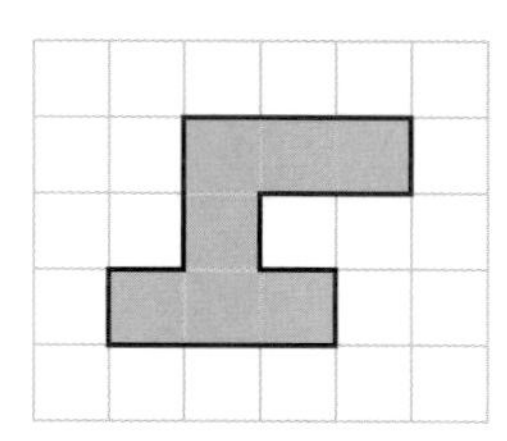

D

3 A shape has a perimeter of 12 cm.
It is enlarged by a scale factor of 3.
What is the perimeter of the enlargement?

4 **a** Copy the shape and the point O onto squared paper.

b Enlarge the shape by a scale factor of 2 using O as the centre of enlargement.

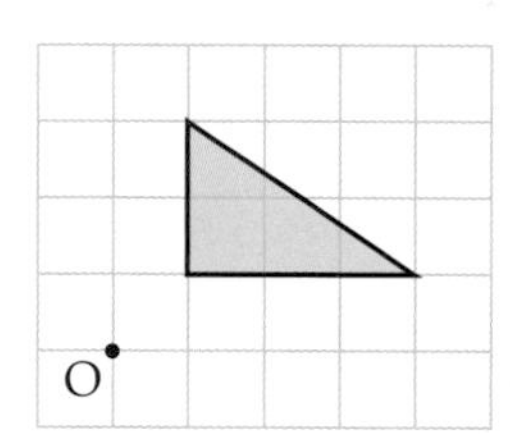

5 In this diagram, shape B is an enlargement of shape A.

a What is the scale factor of the enlargement?

b What are the coordinates of the centre of enlargement?

Copy the shapes then draw in the rays.

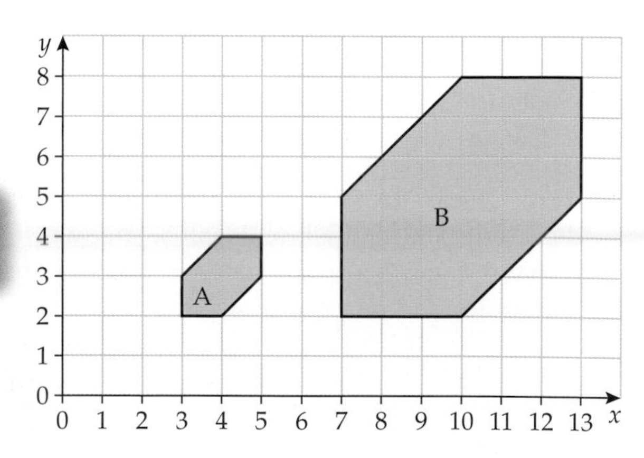

6 **a** Copy the shape and the point O onto squared paper.

b Enlarge the shape by a scale factor of 2 using O as the centre of enlargement.

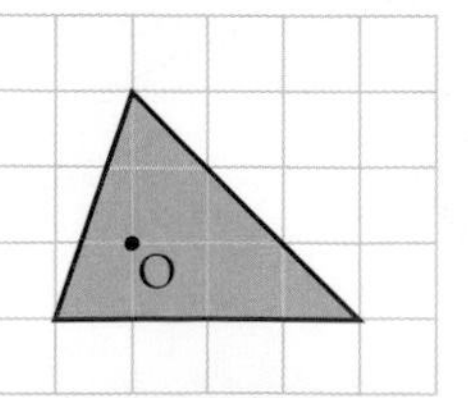

Trial and improvement

Links to:
Foundation Student Book
Ch34, pp. 536–541

Key Points

Using a calculator

- Calculators use the correct order of operations.
- You need to know what these calculator keys do:

(−) Enter a negative number.

x^2 Square a number.

x^3 Cube a number.

√■ Find the square root of a number.

D

Enter a fraction. Use the down arrow to enter the bottom of the fraction.

S⇔D Change the answer from a fraction or square root to a decimal.

Trial and improvement

C

- Some equations cannot be solved using algebra.
- Use trial and improvement to find solutions to these equations. Substitute a value of x into the equation and see how close it is to the value you want. Try again with a different value. The more values you try, the closer you can get to the solution.

34.1 Using a calculator

D

Use a calculator to work out the answers to all the questions in this exercise.
For each answer, write down all the digits on your calculator display, then round your answer to 1 d.p.

1 Work out the value of the following.

a 7.82^2 **b** $\sqrt{782}$ **c** $7.82^3 - 7.82$

d $\frac{(1.2 + 3.4)^2}{5.6}$ **e** $\sqrt{12.34 \times 5.67}$ **f** $(12.34 - 5.678)^2$

2 Work out the value of $x^3 - 10x$ for the following values of x.

a $x = 5$ **b** $x = 5.1$ **c** $x = 0.5$ **d** $x = -0.51$

34.2 Trial and improvement

C

1 Solve the equation $x^3 + x^2 = 42$ using trial and improvement.
Copy the table, and use it to help you solve the equation.
Give your answer correct to 1 d.p.

x	$x^3 + x^2$	Comment
3	36	too low
4	80	too high

Add as many rows to the table as you need.

2 Use trial and improvement to solve the equation $5x^3 + 2x = -850$.
Give your answer correct to 1 d.p.

3 The equation $\frac{2x^3 + 10}{x^2} = 10$ has a solution between 3 and 6.

a Find this solution using trial and improvement.
Give your answer correct to 1 d.p.

b This equation has another solution between 0 and −2.
Find this solution using trial and improvement.
Give your answer correct to 1 d.p.

Quadratic graphs

Links to:
Foundation Student Book
Ch35, pp. 542–555

Key Points

Quadratic graphs

D C

A quadratic function has a term in x^2. It may also have a term in x and a number. It does not have any terms with powers of x higher than 2.

The graph of a quadratic function is a curve in the shape of a U.

All quadratic graphs are symmetrical about a line parallel to the y-axis.

The line of symmetry is given as 'x = a number'.

Graphs can be used to solve quadratic equations. For example, solving $x^2 - 2x - 1 = 0$ means looking to see where the curve crosses the x-axis ($y = 0$).

Solving $x^2 - 2x - 1 = 6$ means looking to see where the curves cross the line $y = 6$.

The points of intersection of the graphs of a quadratic function and a linear function can be found by plotting the graphs on the same axes.

35.1 Graphs of quadratic functions

D

1 a Copy and complete the table of values for $y = x^2 + 4$.

x	−3	−2	−1	0	1	2	3
x^2	9					4	
+4	+4	+4	+4	+4	+4	+4	+4
$y = x^2 + 4$	13					8	

Draw your y-axis going from 0 to 13 and your x-axis going from −3 to + 3.

b Draw the graph of $y = x^2 + 4$ for values of x from −3 to +3.

2 a Copy and complete the table of values for $y = x^2 - 4$.

x	−3	−2	−1	0	1	2	3
x^2	9						
−4	−4						
$y = x^2 - 4$	5					0	

b Draw the graph of $y = x^2 - 4$ for values of x from −3 to +3.

C

3 a Copy and complete the table of values for $y = x^2 - 6x + 8$.

x	−1	0	1	2	3	4	5	6	7
x^2	1					16		36	
$-6x$	+6					−24		−36	
+8	+8					+8		+8	
$y = x^2 - 6x + 8$	15					0		8	

b Draw the graph of $y = x^2 - 6x + 8$ for values of x from −1 to +7.

c What are the coordinates of the lowest point?

d Write down the line of symmetry.

4 Viktor throws a ball vertically upwards.
The height of the ball above the ground after t seconds is given by the function
$h = 12t - 2t^2$
where h is the height, in metres, of the ball above the ground.

a Copy and complete the table of values for $h = 12t - 2t^2$.

t	0	1	2	3	4	5	6
$12t$			24			60	
$-2t^2$			−8			−50	
$h = 12t - 2t^2$			16			10	

b Draw the graph of $h = 12t - 2t^2$.
c Use your graph to find the height of the ball after 4 seconds.
d How long does it take the ball to reach its maximum height?
e What is the maximum height of the ball?
f What is the total length of time that the ball is in the air?

Plot the t-values along the x-axis and the h-values along the y-axis.

35.2 Solving quadratic equations graphically

1 a Copy and complete the table of values for $y = x^2 - 5$.

x	−3	−2	−1	0	1	2	3
y	4		−4			−1	

b Draw the graph of $y = x^2 - 5$.
c Use your graph to find the solutions of the equation $x^2 - 5 = 0$.
d Draw the line $y = 3$ on your graph.
Write down the x-coordinates of the points where the line $y = 3$ crosses the curve $y = x^2 - 5$.
e Write down the quadratic equation whose solutions are the answers to part d.

Draw your y-axis going from −5 to 5 and your x-axis going from −3 to + 3.

Write down the x-values where your graph crosses the x-axis.

2 a Copy and complete the table of values for $y = x^2 - 2x - 1$.

x	−3	−2	−1	0	1	2
y	14	7			−2	

b Draw the graph of $y = x^2 - 2x - 1$.
c Use your graph to find the solutions of the equa
d Draw the line $y = 5$ on your graph.
Write down the coordinates of the points where cross.
e Show that the solutions to the quadratic equatior found at this point.

3 a Draw the graph of $y = x^2 - 4x - 2$ for values of x
b Use your graph to solve
i $x^2 - 4x - 2 = 0$
ii $x^2 - 4x - 2 = 7$.
c Can the quadratic $x^2 - 4x - 2 = -7$ be solved?
Explain your answer.

Links to:
Foundation Student Book
Ch36, pp. 556–568

Key Points

Constructions

C

Constructions must be drawn using *only* a straight edge (ruler) and a pair of compasses.

Leave all construction lines and arcs on the diagram as evidence you have used the correct method.

Perpendicular bisector of a line segment

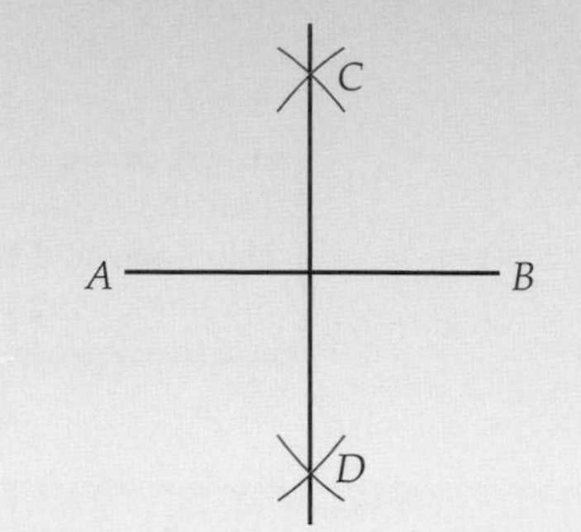

The bisector of an angle

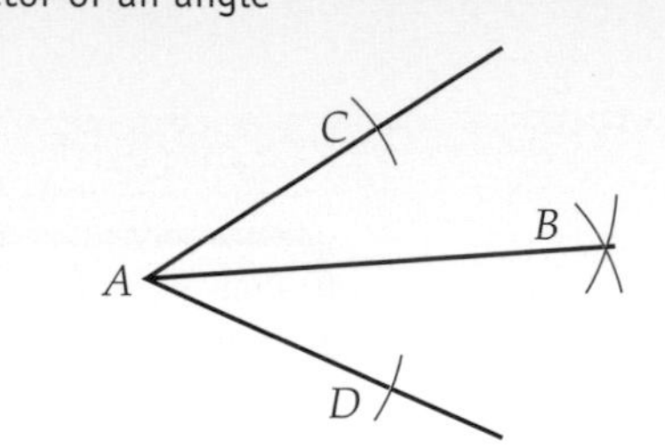

Locus

C

The locus of points that are the same distance from a fixed point is a circle.

The locus of points that are the same distance from a fixed line is two parallel lines, one each side of the given line.

The locus of points that are the same distance from a fixed line segment *AB* is a 'racetrack' shape. The shape has two lines parallel to *AB* and two semicircular ends.

The locus of points that are the same distance from two fixed points is the perpendicular bisector of the line segment joining the two points.

The locus of points that are the same distance from two fixed lines is the bisector of the angle formed by the lines.

36.1 Constructions

C

1 a Draw a line segment, XY, about 6 cm long.

b Construct the perpendicular bisector of the line segment XY.

c Check the accuracy of your construction by measuring with a ruler.

2 a Using a ruler, draw a rough sketch of this angle.

…he bisector of angle CAB.

…racy of your construction by measuring the angle with a protractor.

…ius 5 cm.

…your circle so that all three

…ch the circumference of the circle.

…bisector of each side of

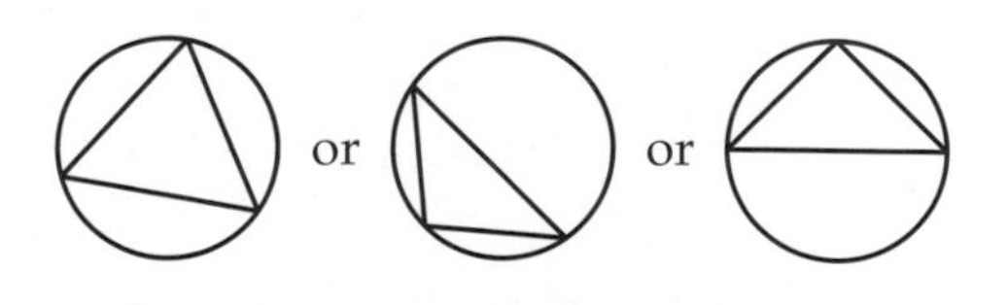

C

4 a Draw a right-angled triangle with sides 6 cm, 8 cm and 10 cm.

b Construct the perpendicular bisector of each side of your triangle.

c Put your compass on the point where your bisectors cross, open them out to point C and draw a circle.
What do you notice?

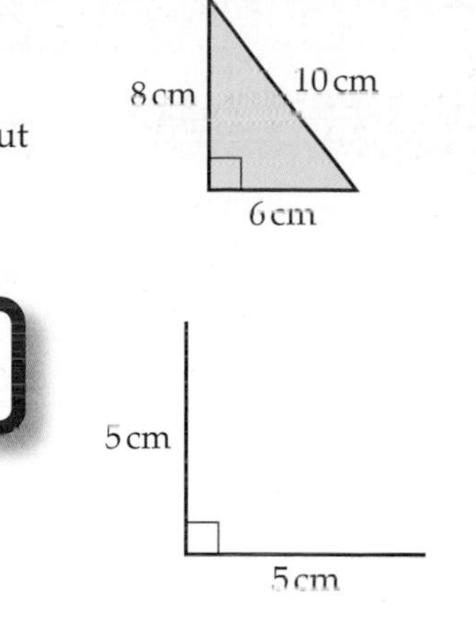

5 a Make an accurate copy of this diagram.

b Construct the angle bisector of the right angle.

c Bisect one of your new angles.

d Measure one of the angles you have just created.

e Work out the size of the angle you have just measured.
Was your measurement accurate enough?

> **Four equal angles make 90°.**

36.2 Locus

C

1 The line XY is 8 cm long. X ——————— Y
Make a copy of the line XY.
Draw the locus of the points that are exactly 2.5 cm from XY.

2 a Using a ruler, sketch this shape.

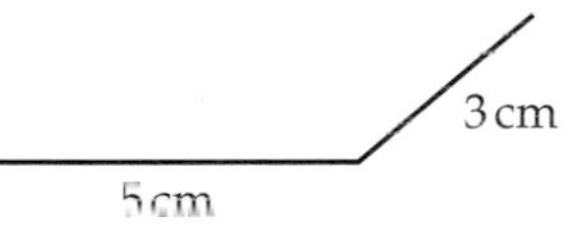

b Draw the locus of points that are 3 cm from the line.

3 $ABCD$ is a square field.
There is a fence around the perimeter of the field.
One goat is tethered by a rope to corner B.
Another goat is tethered by a rope to corner C.
When the ropes are tight, the goats can reach corner D.

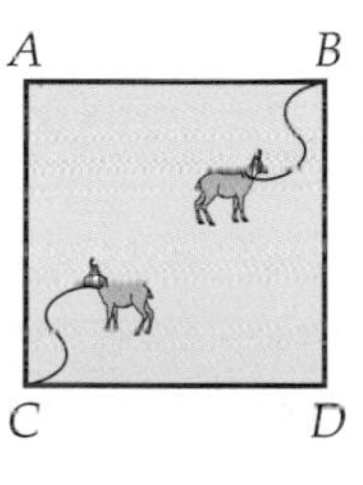

a Draw a square to represent the field.

b Show the area of grass that both goats can reach.

4 a Draw an equilateral triangle ABC of side length 5 cm.

b Construct the locus of points that are the same distance from AB as they are from BC.

c Draw the locus of points that are exactly 3 cm from A.

d Shade the region inside the triangle that is closer to AB than to BC and more than 3 cm from A.

C

FUNCTIONAL

5 A company wants to build an out-of-town shopping centre.
The customers will mainly come from three towns, shown as A, B and C on the diagram.
Planning permission will be given if the centre is

- within 5 miles of town B and within 6 miles of town A
- closer to road AB than to road BC.

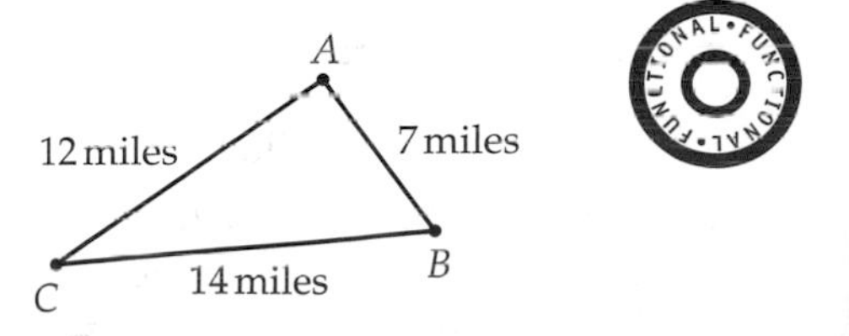

a Make a copy of the diagram using a scale of 1 cm = 1 mile.

b Construct the region in which the company can build the shopping centre.

A03

Links to:
Foundation Student Book
Ch37, pp. 569–583

Key Points

Pythagoras' theorem

C

In a right-angled triangle the longest side is called the hypotenuse.

The hypotenuse is always opposite the right angle.

For a right-angled triangle with sides of length a, b and c, where c is the hypotenuse, Pythagoras' theorem states that $a^2 + b^2 = c^2$.

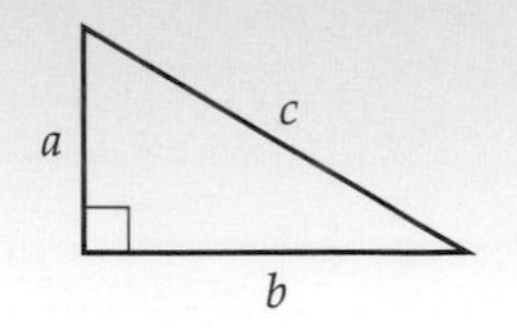

You can calculate the lengths of the shorter side of a right-angled triangle using

- $a^2 = c^2 - b^2$
- $b^2 = c^2 - a^2$

Length of a line segment

C

Pythagoras' theorem can be used to find the length of a diagonal line AB, given the coordinates of A and B.

37.1 Pythagoras' theorem

C

1 Write down the letter that represents the hypotenuse in this triangle.

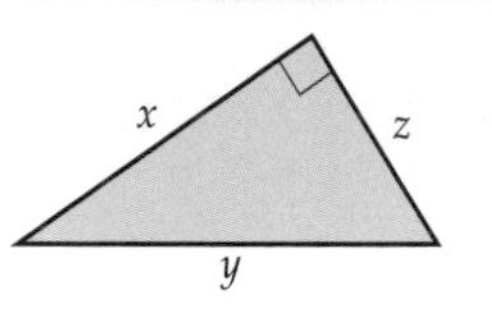

2 For this triangle:

a write down the letter that represents the hypotenuse

b write down the formula for Pythagoras' theorem.

p r q

37.2 Finding the hypotenuse

C

1 Calculate the length of the hypotenuse in this triangle.

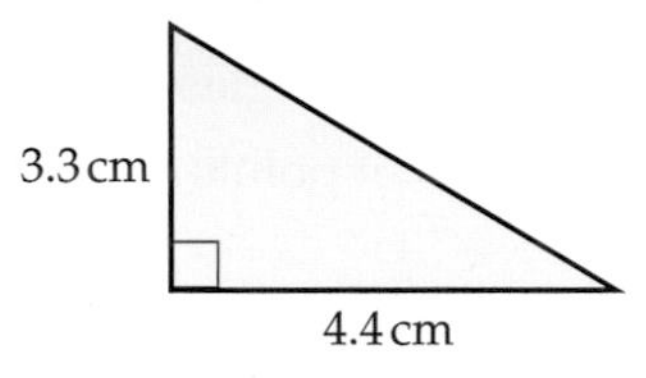

2 A boat sails 20 km due north then 50 km due east.
How far is the boat from its starting point?
Give your answer to the nearest km.

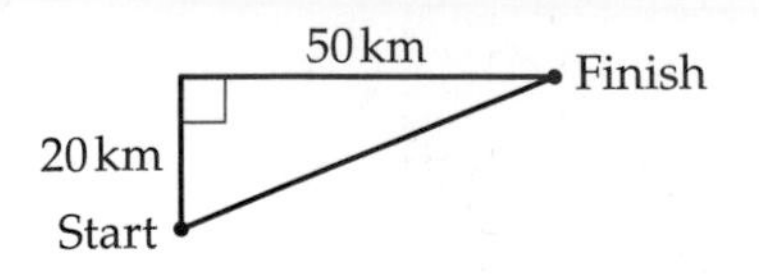

3 A rectangular piece of A4 paper measures 297 mm by 210 mm.
What is the length of the diagonal on a sheet of A4 paper?
Give your answer to the nearest millimetre.

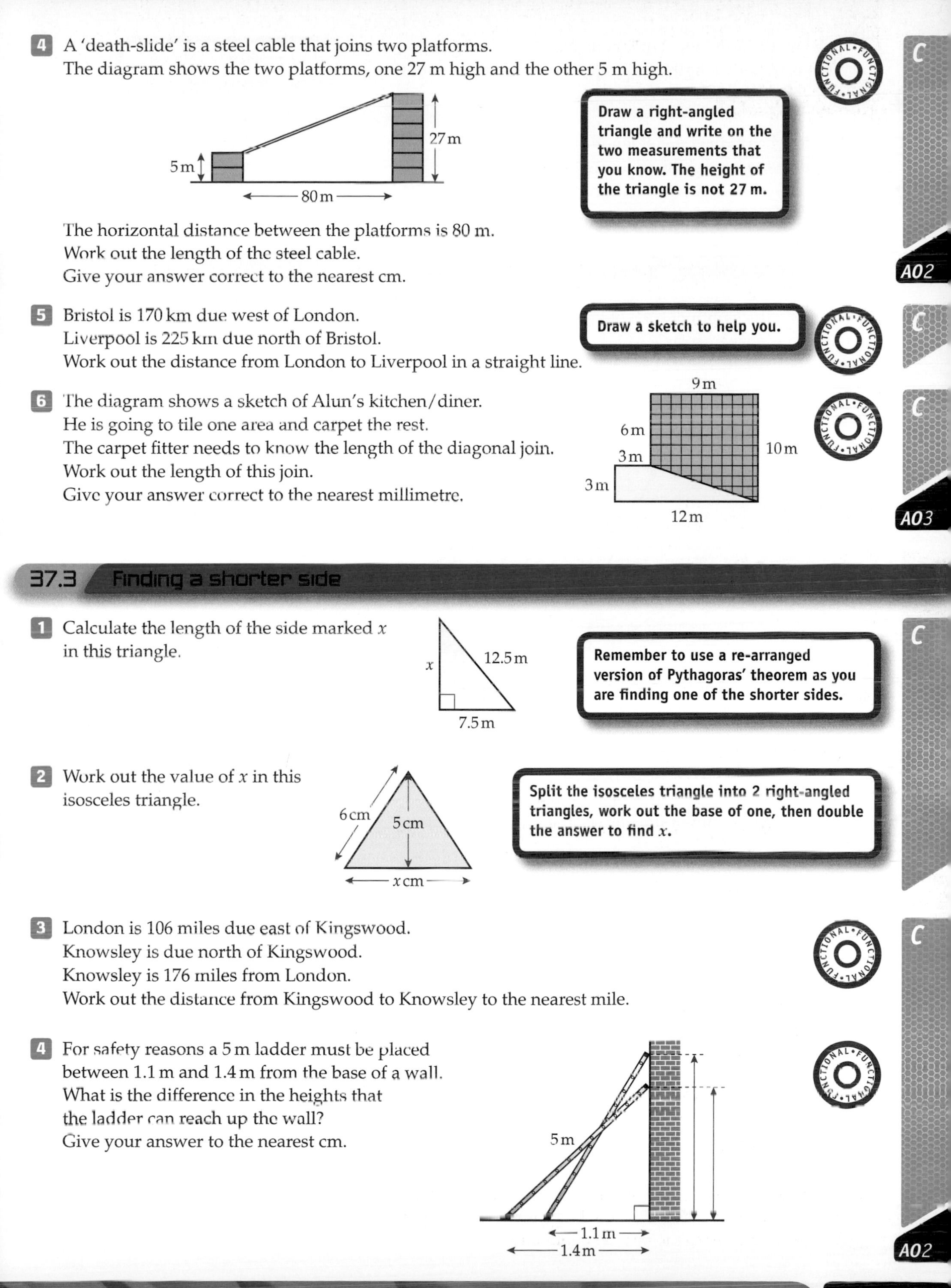

4 A 'death-slide' is a steel cable that joins two platforms.
The diagram shows the two platforms, one 27 m high and the other 5 m high.

Draw a right-angled triangle and write on the two measurements that you know. The height of the triangle is not 27 m.

The horizontal distance between the platforms is 80 m.
Work out the length of the steel cable.
Give your answer correct to the nearest cm.

C
AO2

5 Bristol is 170 km due west of London.
Liverpool is 225 km due north of Bristol.
Work out the distance from London to Liverpool in a straight line.

Draw a sketch to help you.

C

6 The diagram shows a sketch of Alun's kitchen/diner.
He is going to tile one area and carpet the rest.
The carpet fitter needs to know the length of the diagonal join.
Work out the length of this join.
Give your answer correct to the nearest millimetre.

C
AO3

37.3 Finding a shorter side

1 Calculate the length of the side marked x in this triangle.

Remember to use a re-arranged version of Pythagoras' theorem as you are finding one of the shorter sides.

C

2 Work out the value of x in this isosceles triangle.

Split the isosceles triangle into 2 right-angled triangles, work out the base of one, then double the answer to find x.

3 London is 106 miles due east of Kingswood.
Knowsley is due north of Kingswood.
Knowsley is 176 miles from London.
Work out the distance from Kingswood to Knowsley to the nearest mile.

C

4 For safety reasons a 5 m ladder must be placed between 1.1 m and 1.4 m from the base of a wall.
What is the difference in the heights that the ladder can reach up the wall?
Give your answer to the nearest cm.

AO2

C

AO2

5 A square has a diagonal of length 10 cm.
What is the side length of the square to the nearest mm?

Draw a sketch to help you.

C

AO3

6 A square has a diagonal of length 12 cm.
Calculate the area of the square.

37.4 Calculating the length of a line segment

C

1 Calculate the length of each of these lines.
Give your answer correct to 1 d.p. where appropriate.

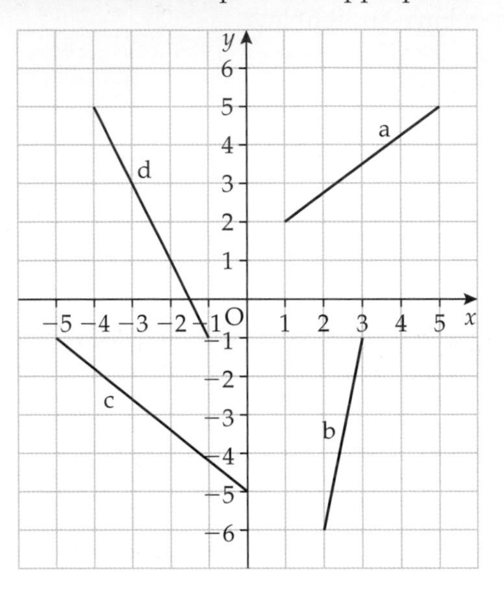

C

AO2

2 Calculate the length of the line segment from (4, 1) to (9, 13).

Draw a sketch to help you.

3 Calculate the length of the line segment from (−3, 2) to (3, −2) correct to 1 d.p.

EXAM PRACTICE PAPERS

Unit 1 Foundation Statistics and Number

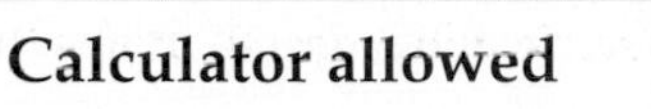

1 Twenty students were asked which mobile phone provider they used.
Here are the results

Orange	Vodafone	O_2	Vodafone	Orange	O_2	Vodafone
Vodafone	Virgin	Vodafone	Orange	Vodafone	Orange	Vodafone
Vodafone	Orange	O_2	Virgin	Orange	Vodafone	Virgin

a Copy and complete the table **(2 marks)** | F

Mobile phone provider	Tally	Frequency
Orange		

b Which mobile phone provider was most popular? **(1 mark)** | G | **Funct.**

c Copy and complete the pictogram to show the results. **(2 marks)** | G

Key: ☺ represents 2 people

Orange	
Vodafone	

2 A shop sells hats and scarves which are either red or blue.
These compound bar charts show the number of hats and scarves in stock on 1 December and on the 31 December.

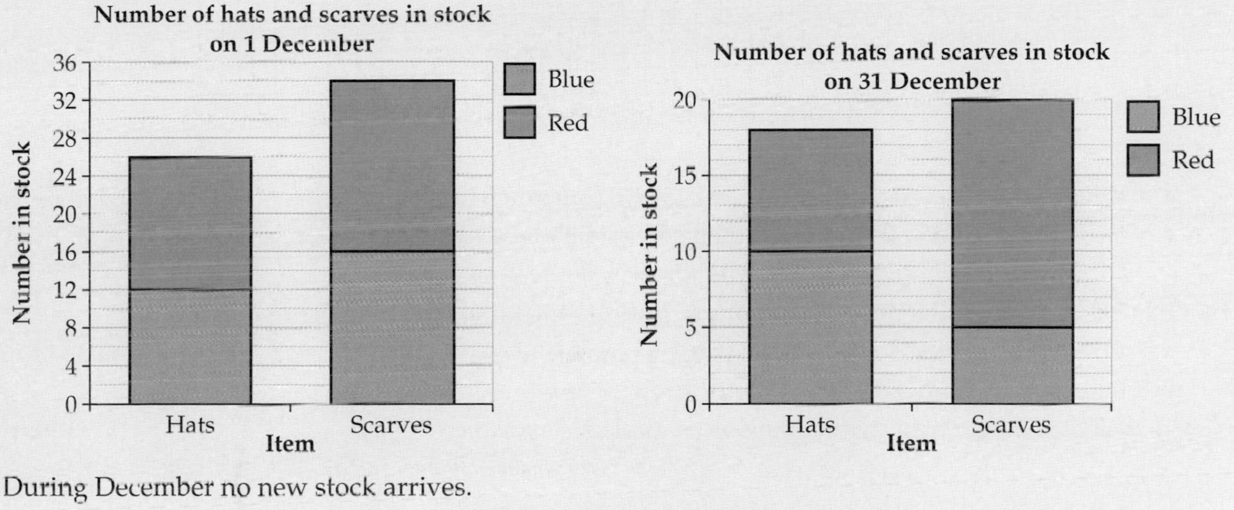

During December no new stock arrives.
Complete a two-way table to show the number of hats and scarves sold in December. **(3 marks)** G | AO3

3* These are the prices that Harry the Electrician charges for a job.

Normal callout: **£30 per hour**
Emergency callout: **£35 per hour + £25 fixed charge**

a On Tuesday morning Harry goes to an emergency callout. The job takes him two hours.
How much does he charge the customer? **(2 marks)** | G | **Funct.**

b On Tuesday afternoon Harry goes to a normal callout. He charges the customer £90.
How long did the job take? **(2 marks)** | G | **Funct.**

c Mr Patel makes an emergency call to Harry. He needs an electrician for about 4 hours.
Another electrician will charge Mr Patel £150 for the same job.
Is it cheaper for Mr Patel to use Harry or the other electrician?
You must show all your working. **(3 marks)** D | AO3 | **Funct.**

4 Mr Harris is planning a two-week holiday to Capetown in South Africa.
The vertical line graph shows the monthly average rainfall in Capetown.

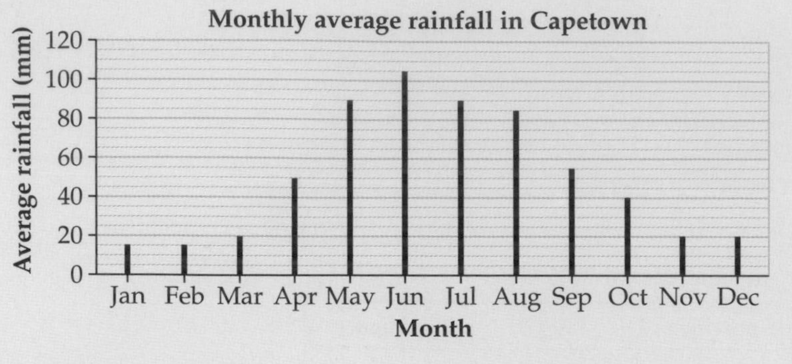

a What is the average rainfall in April? **(1 mark)** | G | **Funct.**

b Which month has the most rain? **(1 mark)** | G | **Funct.**

c Work out the range of this data. **(2 marks)** | F | **Funct.**

d Mr Harris can only go only holiday in July, August, September or October.
During which of these months do you think he should go on holiday?
Give a reason for your answer. **(1 mark)** | G | **AO2** | **Funct.**

This bar chart shows the average monthly daytime temperatures in Capetown.

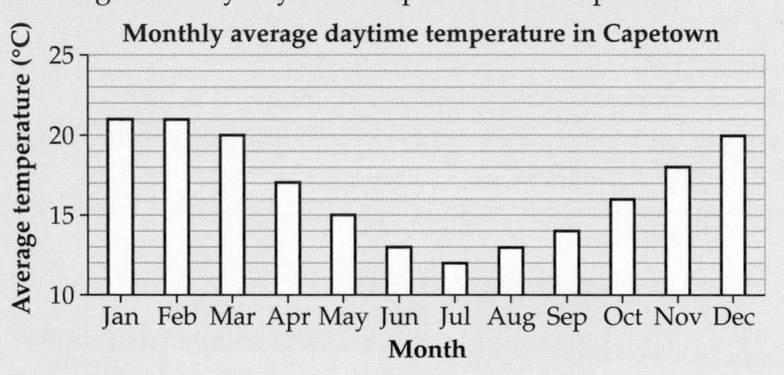

e What is the average daytime temperature in Capetown in August? **(1 mark)** | G | **Funct.**

f Do you think that Mr Harries should still go on holiday in the month you gave as your answer to part **d**? Explain your answer. **(1 mark)** | G | **AO2** | **Funct.**

5 In a cafe you can order apple pie or plum crumble with either cream or ice cream.

a Write down all the combinations that are possible to order. **(1 mark)** | F | **AO2**

b One person is selected at random.
What is the probability that they order apple pie and cream? **(1 mark)** | F

6 Alex has these letter cards.

a What fraction of the cards have the letter R?
Give your answer in its simplest form. **(2 marks)** | F

b What fraction of the cards do not have the letter R? **(1 mark)** | E

c What percentage of the cards do not have the letter R? **(1 mark)** | E

Alex shuffles the cards then selects one at random.

d Tick a word from the list below to describe the probability of each of the following.

i Alex selects a card with the letter I.

Impossible Unlikely Evens Likely Certain **(1 mark)** | G

ii Alex selects a card that is not a vowel.

Impossible Unlikely Evens Likely Certain **(1 mark)** | E

7 Carlos has these two spinners.

He spins the spinners at the same time.

He adds together the two numbers the spinners land on to give him a score.

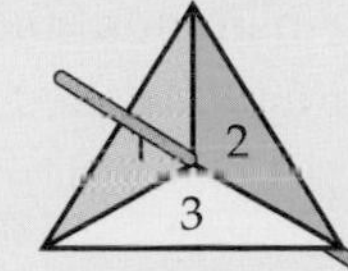

a Complete the table below to show his possible scores. **(2 marks)** | F

+	1	2	3	4	5
1	2				6
2					
3	4				

b Work out the probability Carlos gets a score of 4. **(1 mark)** | D

c Work out the probability that his score is an odd number. **(2 marks)** | D

8 From May to August Mandy runs boat trips to go dolphin watching.

The table shows the number of trips each month when dolphins are seen.

It also shows the angles she has worked out for a pie chart to show this information.

Month	Number of trips	Angle in pie chart
May	8	40°
June	15	75°
July	35	175°
August	32	160°

a Explain how you know that the angles for the pie chart cannot all be correct? **(1 mark)** | E | **AO2** | **Funct.**

b Draw a pie chart to represent the data. **(4 marks)** | E | **AO2** | **Funct.**

9 **a** The stem and leaf diagram shows the weights of the players in a rugby team.

7	4 5
8	6 8 9 9
9	0 3 4 5 8 9
10	2 3 5

Key: 7 | 4 means 74 kg

i Dafydd says 'The mean weight of the players in the team is 93 kg.'
Is Dafydd correct? Explain your answer. **(2 marks)** | E | **AO2** | **Funct.**

ii During a match, the 75 kg player is injured and is replaced by a player that weighs 82 kg.
Will the mean weight of the players in the team increase, decrease or stay the same?
Explain your answer. **(1 mark)** | E | **AO2** | **Funct.**

b The group frequency table shows the weights of the players in a different rugby team.

Weight, w kg	Frequency
$65 \leq w < 75$	
$75 \leq w < 85$	5
$85 \leq w < 95$	5
$95 \leq w < 105$	
Total	**15**

The mean weight of the players is 84 kg.

Complete the frequency column in the table above.

You must show your workings. **(4 marks)** | C | **AO3**

10 Mair wants to find out the ages of people that play videogames.
She has written this question for her survey.
In what age group are you? Please tick one box only.
20 to 40 years ☐ 40 to 60 years ☐ 60 to 80 years ☐

a Write down two criticisms of the question. **(2 marks)** | D | AO2 | **Funct.**

b Re-write the question to make it more suitable. **(2 marks)** | C | AO2 | **Funct.**

11* A bicycle normally costs £800.
It is reduced in a sale by 20%.
It doesn't sell, so two weeks later it is reduced by a further 25%.
Show that overall the bicycle has been reduced in price by 40%. **(3 marks)** | C | AO2 | **Funct.**

Unit 2 Foundation Number and Algebra — Non-calculator

1 Here are some number cards. 3 10 4 12 20 30 2 5

a Which two cards have numbers that are multiples of 6?
.... and .... **(2 marks)** | G

b Which two cards have numbers that are factors of 15?
.... and .... **(2 marks)** | G

2 Sharon runs dog-training courses.
The courses are in obedience or agility.
Each course lasts eight weeks.
The table shows the prices for an eight-week course.

Course	Price for eight weeks
Obedience	£48
Agility	£36

a How much does it cost per week to do the agility course? **(1 mark)** | G

Sharon has six dogs on her obedience course and seven on her agility course.

b How much money does Sharon earn from her obedience course? **(1 mark)** | G

c How much money does Sharon earn altogether from these two courses? **(2 marks)** | G

3 Sham borrows £7100 from his Dad.
He pays his Dad back over 12 months.
He works out how much he must pay his Dad each month.
His answer is shown on the calculator.

591.6666667

a Round his answer to the nearest £100. **(1 mark)** | G

b Round his answer to two decimal places. **(1 mark)** | F

4 Here is a number pattern.
$50 - 2 = 48$
$50 - 2 - 4 = 44$
$50 - 2 - 4 - 6 = 38$

a Write down the next two lines of the pattern. **(4 marks)** | G

b Describe in words the rule for continuing the sequence 48, 44, 38, ... **(1 mark)** | F

5 Work out

a $136.2 + 18.9$ **(1 mark)** | G

b $12.8 - 9.2$ **(1 mark)** | G

c $344 \div 8$ **(1 mark)** | G

d 128×37 **(3 marks)** | F

6 Here is a formula to work out the time it takes to cook a piece of beef.

Time in minutes = weight of beef in kg $\times$ 40 + 30

a How many minutes does it take to cook a piece of beef that weighs 2 kg? **(2 marks)** | G | **Funct.**

b What is the weight of piece of beef that takes 3 hours 10 minutes to cook? **(2 marks)** | F | **Funct.**

7* Here are the rules for a school quiz.

> All contestants answer the same **five questions**.
>
> A contestant scores **five points** if they are the only person to get the answer right.
>
> A contestant scores **three points** if they get the answer right, but they aren't the only one.
>
> A contestant scores **no points** if they get the answer wrong.

Moira and Sanjay took part in the quiz.

Moira scored 12 points and Sanjay scored 13 points.

Who had the most questions right?

You must show your working. **(4 marks)** | F | **AO3**

8 Work out the value of:

a $6q$ when $q = 5$ **(1 mark)** | F

b $\frac{1}{2}x + 2y$ when $x = 14$ and $y = -4$. **(2 marks)** | E

9 Helen has three children, Alex, Billy and Charlie.

Alex has £120 in his bank account.

Billy has £105 in his bank account.

Altogether the children have £400 in their bank accounts.

a How much does Charlie have in his bank account? **(2 marks)** | F

Helen pays a total of £140 into their bank accounts.

They now all have the same amount of money.

b How much did Helen pay into Billy's bank account? **(3 marks)** | E | **AO2**

10 a Simplify the expression $3t + 4v - 2t + v$ **(2 marks)** | E

b Multiply out the brackets in this expression $x(2x + 5)$ **(2 marks)** | D

11* Rhys says '$\frac{2}{3}$ is smaller than $\frac{5}{8}$'.

Is Rhys correct?

You must show working to support your answer. **(3 marks)** | E | **AO2**

12 When Ian goes on holiday, the exchange rate from British pounds to American dollars is £1 = \$1.60

a Copy and complete this conversion table between pounds (£) and dollars (\$).

£	0	10	50
\$		16	

(1 mark) | E

b On graph paper draw a conversion graph between pounds (£) and dollars (\$).

Plot pounds on the x-axis, from £0 to £50, and dollars on the y-axis, from \$0 to £80. **(1 mark)** | E

c When Ian is on holiday he buys a jumper for \$40.

Use your graph to work out how much the jumper costs in pounds? **(1 mark)** | E | **Funct.**

13 Last year Shani worked 30 hours a week and was paid £9 per hour.

This year Shani's hours were reduced by 10% but her pay was increased by 11%.

Is Shani now earning more or less per week than she did last year?

You must show your working. **(4 marks)** | D | **AO2** | **Funct.**

14 The nth term of a sequence is $6n + 2$.

Show that all the terms in the sequence are even. **(2 marks)** | D | **AO2**

15 The formula to convert a temperature in degrees Celsius (°C) to degrees Fahrenheit (°F) is:

$F = 1.8C + 32$ F is the temperature in degrees Fahrenheit (°F)
C is the temperature in degrees Celsius (°C)

a Jerry is going on holiday to Spain.
The temperature forecast is 40°C.
What is this temperature in °F? **(2 marks)** | D

b Anil is going on holiday to America.
The temperature forecast is 86°F.
What is this temperature in °C? **(2 marks)** | C | **AO2**

16 Use approximations to estimate the value of $\frac{207.43 + 48.5}{5.293}$ **(3 marks)** | C

17 Caz runs her own business from home.
At the end of the year she works out her profit by subtracting her total expenses from her total income.
On average her income is £2 200 per month.
This table gives information on her expenses.

Item	**Household bill**	**Amount allowed for expenses**
Internet costs	£15 per month	$\frac{2}{3}$ of total
Telephone	£32 per month	75% of total
Electricity	£280 per quarter (3 months)	20% of total
Stationary/software	£340 per year	100% of total

Calculate Caz's profit for this year. **(6 marks)** | C | **AO3 Funct.**

Unit 3 Foundation Geometry and Algebra — Calculator allowed

1 a Shona is facing south. She makes a $\frac{1}{4}$ turn clockwise.
In which direction is she now facing? **(1 mark)** | G

b How many degrees are there in a $\frac{1}{4}$ turn clockwise? **(1 mark)** | G

c This diagram shows a seven sided shape.

i How many acute angles does the shape have? **(1 mark)** | G

ii What type of angle is angle x? **(1 mark)** | G

iii What type of angle is angle t? **(1 mark)** | F

2 The diagram shows a circle.
Fill in the missing words in these sentences.

a The line AB is called the .... **(1 mark)** | G

b The curved line BC is called an .... **(1 mark)** | G

c The shaded area is called a .... **(1 mark)** | G

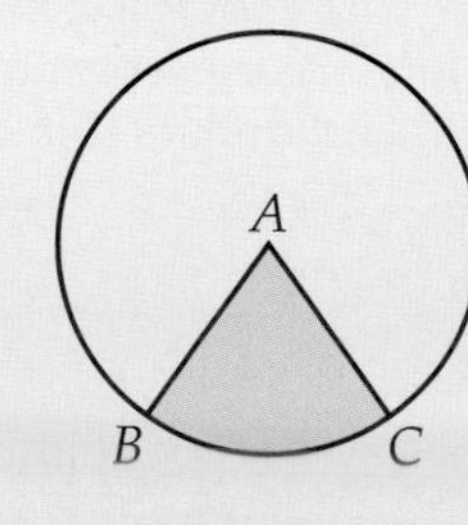

3* Ted has a maximum of £400 to spend on a holiday.
He needs to pay for a hotel, flights and insurance.
He sees this advert.
Can Ted afford this holiday?
You **must** show your working.

Holiday to Spain - Special Offer!
Hotel: £287 per person
Flights: £125 per person
Insurance: £19 per person
Buy online and get 10% off total price.

(2 marks) | G | **AO1 (2 marks)** | F | **AO2 Funct.**

4 a What name is given to this type of triangle?

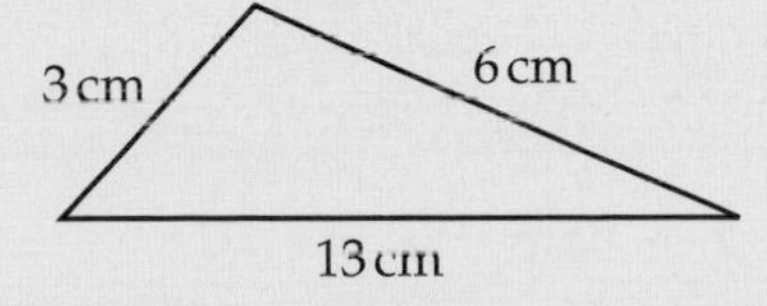

(1 mark) | G

b Copy the following triangles and draw on all their lines of symmetry.

i

ii

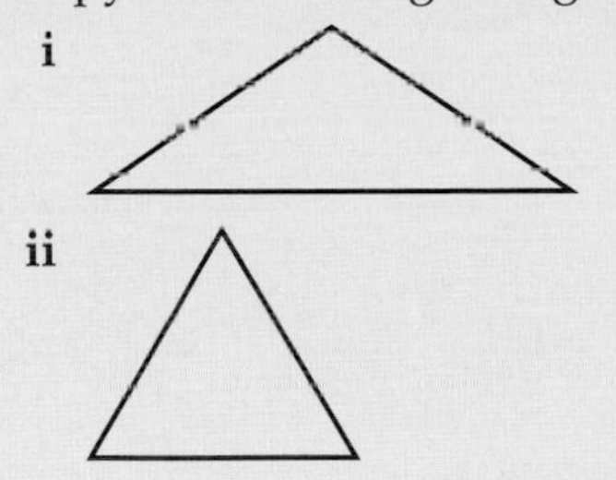

(1 mark) | G

(2 marks) | G

c Draw a rectangle 5 cm long and 2 cm wide. By drawing three extra lines, show how you can divide the rectangle into four right-angled triangles. **(2 marks)** | G

5 Write down the amount shown on each of these scales.

a

b

(1 mark) | G | **Funct.**

(1 mark) | G | **Funct.**

c Write down the two different times that clock could be showing using the 24-hour clock.

(2 marks) | G | **Funct.**

6 Which two of the following could be the net of a cube?

A B C D

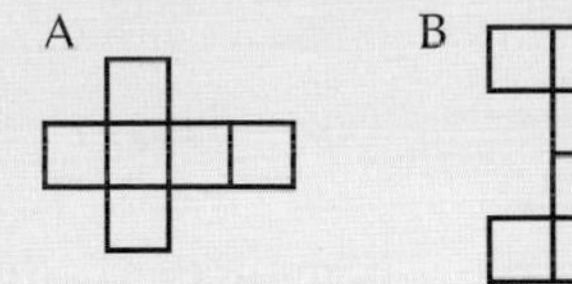

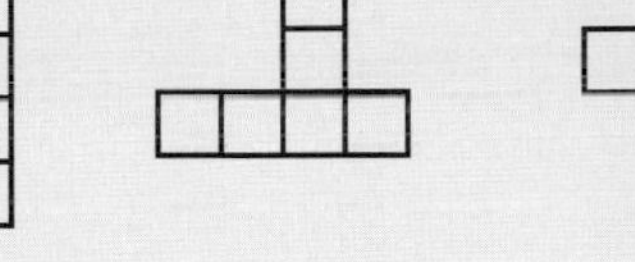

(2 marks) | G

7 Sascha is going buy pet insurance for his dog.
There are two ways of paying.

Pet Insurance	
Monthly payment	£4.55
Yearly payment	£47.50

How much would Sascha save by making a single yearly payment? **(3 marks)** | G | **AO2** | **Funct.**

8 a Here is a quadrilateral.

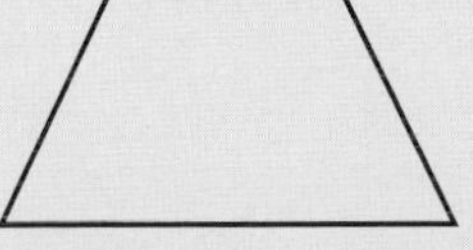

i Write down the name of this quadrilateral. **(1 mark)** | F

ii Write down the order of rotational symmetry of this quadrilateral. **(1 mark)** | F

b Here is a different quadrilateral.

i Write down the name of this quadrilateral. **(1 mark) | F**

ii Write down the order of rotational symmetry of this quadrilateral. **(1 mark) | F**

9 Work out the size of angle x in this diagram.

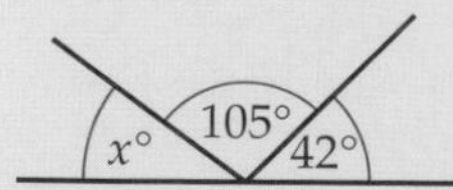

(2 marks) | F

10* This table shows some distances in miles and their equivalent distances in kilometres.

Miles	0	10	20	30	40
Kilometres	0	16	32	48	64

a Draw a graph to show this information.
Plot miles on the x-axis going from 0 to 40, and kilometres on the y-axis going from 0 to 70. **(3 marks) | F**

b Simon drives 36 km.
Use your graph to work out how many miles is this? **(1 mark) | F**

c The distance from Leeds to Chester is 70 miles.
Use your graph to work out the distance from Leeds to Chester in km? **(3 marks) | F | AO2**

11* Sam plots the points A, B and C on the grid shown.

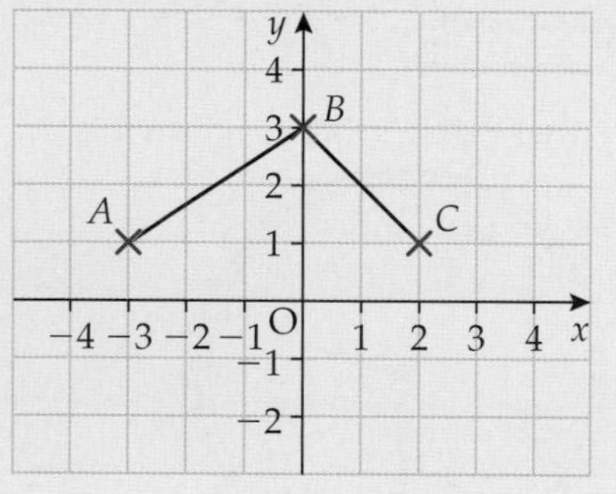

a Sam says 'If I plot point D at $(-1, -1)$ shape $ABCD$ will be a kite'.
Explain why Sam is wrong. **(1 mark) | E | AO2**

b Write down the coordinates of the point D, that would make shape $ABCD$ a kite. **(1 mark) | E | AO2**

12* A bag of sweets contains 60 sweets to one significant figure.
What is the smallest number of sweets that could be in the bag?
Circle your answer from the list below and give a reason for your answer.

59 sweets 54 sweets 55 sweets 57 sweets **(2 marks) | E | AO2**

13 The diagram shows the position of two buoys, A and B.

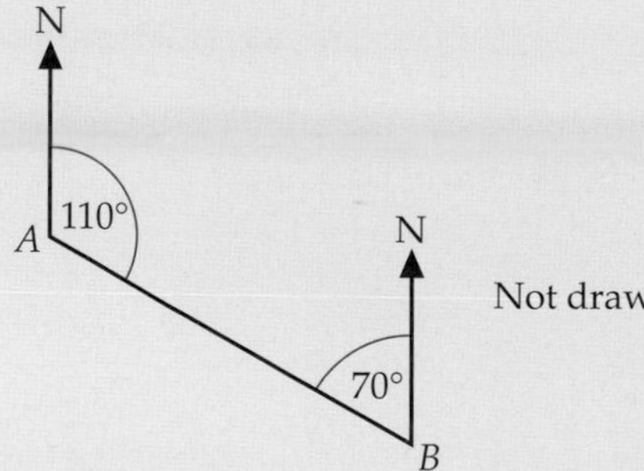

Not drawn accurately

a Write down the bearing of B from A. **(1 mark) | E**

b Work out the bearing of A from B. **(2 marks) | D**

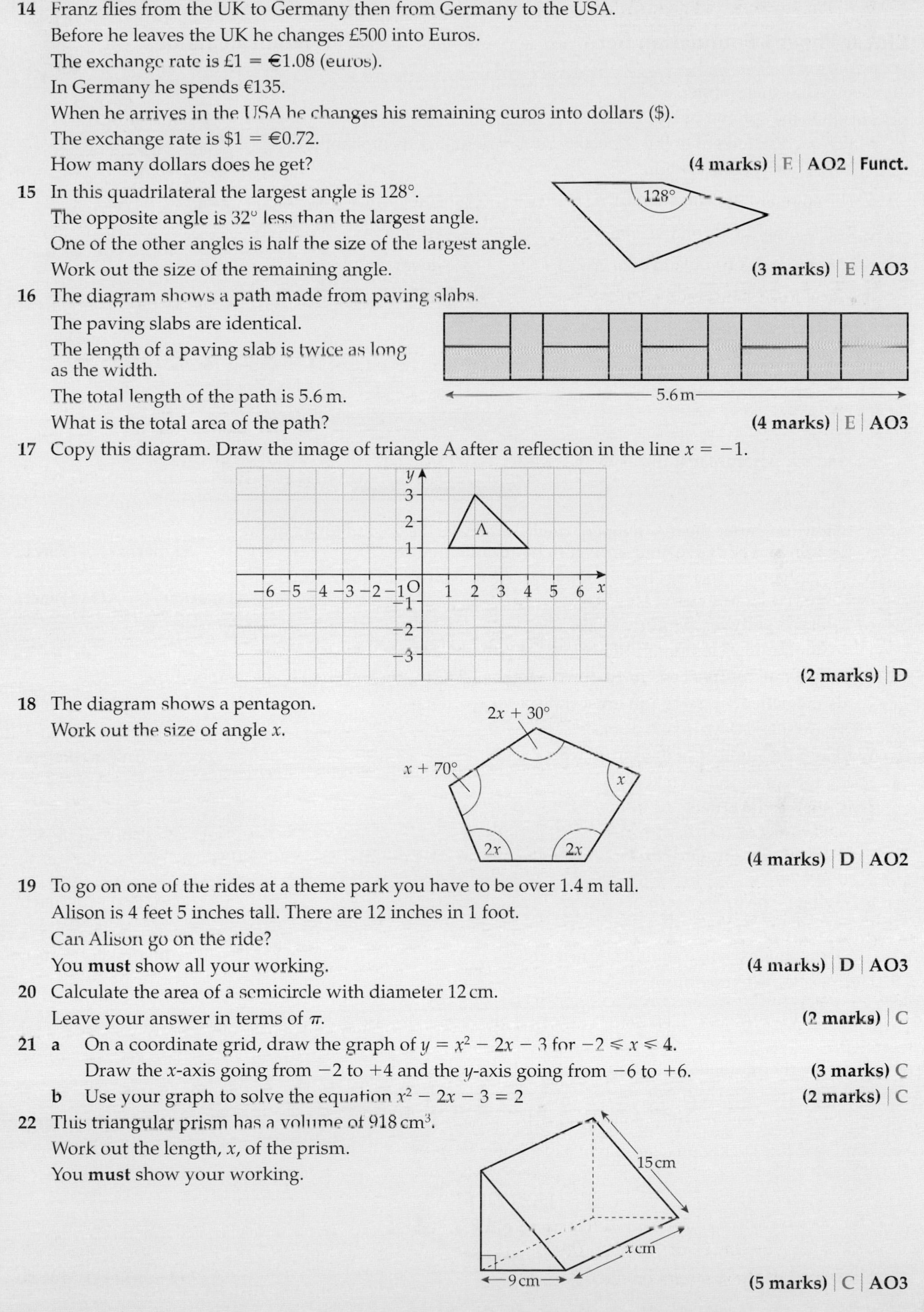

14 Franz flies from the UK to Germany then from Germany to the USA.
Before he leaves the UK he changes £500 into Euros.
The exchange rate is £1 = €1.08 (euros).
In Germany he spends €135.
When he arrives in the USA he changes his remaining euros into dollars ($).
The exchange rate is $1 = €0.72.
How many dollars does he get? **(4 marks)** | E | **AO2** | **Funct.**

15 In this quadrilateral the largest angle is 128°.
The opposite angle is 32° less than the largest angle.
One of the other angles is half the size of the largest angle.
Work out the size of the remaining angle. **(3 marks)** | E | **AO3**

16 The diagram shows a path made from paving slabs.
The paving slabs are identical.
The length of a paving slab is twice as long as the width.
The total length of the path is 5.6 m.
What is the total area of the path? **(4 marks)** | E | **AO3**

17 Copy this diagram. Draw the image of triangle A after a reflection in the line $x = -1$.

(2 marks) | D

18 The diagram shows a pentagon.
Work out the size of angle x.

(4 marks) | D | **AO2**

19 To go on one of the rides at a theme park you have to be over 1.4 m tall.
Alison is 4 feet 5 inches tall. There are 12 inches in 1 foot.
Can Alison go on the ride?
You **must** show all your working. **(4 marks)** | D | **AO3**

20 Calculate the area of a semicircle with diameter 12 cm.
Leave your answer in terms of π. **(2 marks)** | C

21 **a** On a coordinate grid, draw the graph of $y = x^2 - 2x - 3$ for $-2 \leq x \leq 4$.
Draw the x-axis going from −2 to +4 and the y-axis going from −6 to +6. **(3 marks)** C

b Use your graph to solve the equation $x^2 - 2x - 3 = 2$ **(2 marks)** | C

22 This triangular prism has a volume of 918 cm³.
Work out the length, x, of the prism.
You **must** show your working.

(5 marks) | C | **AO3**

1 George has this spinner.
He spins the spinner once.
Match each statement to the correct word that describes its probability.
The first one is done for you.

a The spinner lands on a vowel → likely
b The spinner lands on the letter A
c The spinner doesn't land on a vowel
d The spinner lands on the letter B

impossible
certain
evens
unlikely
likely

(3 marks) | G

2 The thermometers show the temperature of a patient in hospital at 11:30 and at 16:25.

First temperature reading at 11:30.

Second temperature reading at 16:25.

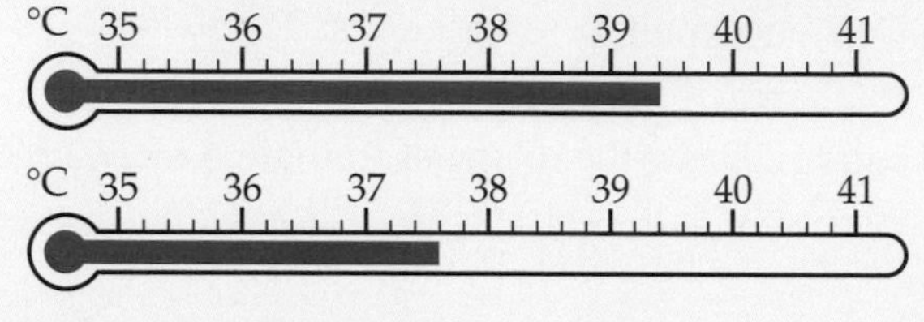

a How long after the first temperature reading was the second temperature reading taken? Give your answer in hours and minutes. **(2 marks)** | G | **Funct.**

b By how many degrees has the temperature of the patient gone down between 11:30 and 16:25? **(3 marks)** | G | **AO2** | **Funct.**

c Normal body temperature is 36.8°C.
After the 16:25 temperature reading the patient says
'My temperature needs to go down another 1.2°C to reach normal body temperature.'
Is the patient correct? You **must** show your working. **(2 marks)** | G | **AO2** | **Funct.**

3 There are seven days in one week.
Work out the number of days in 38 weeks. **(2 marks)** | G

4 Shani goes to an Indian takeaway.
This is what she orders.

Chicken Kurma	£5.20
Special Pilau rice	£1.95
Papadom	55p

a What is the total cost of her order? **(2 marks)** | G | **Funct.**

b Shani pays with a £10 note.
How much change should she receive? **(2 marks)** | G | **Funct.**

5 Here are some temperature cards.

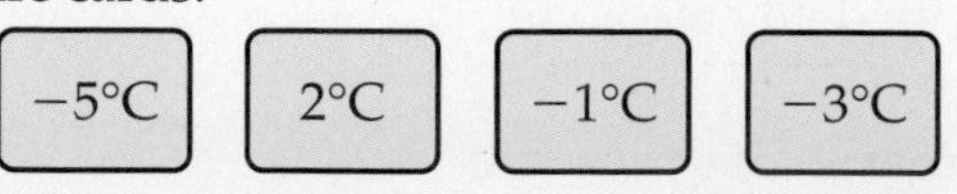

Copy the thermometer below and write on it the correct positions of the temperatures.

(3 marks) | G | **AO3**

6 Hani saw this sticker on a lap-top in a computer shop.

How much is the discount on the lap-top? **(2 marks)** | F | **Funct.**

7 **a** Work out the size of angle x.

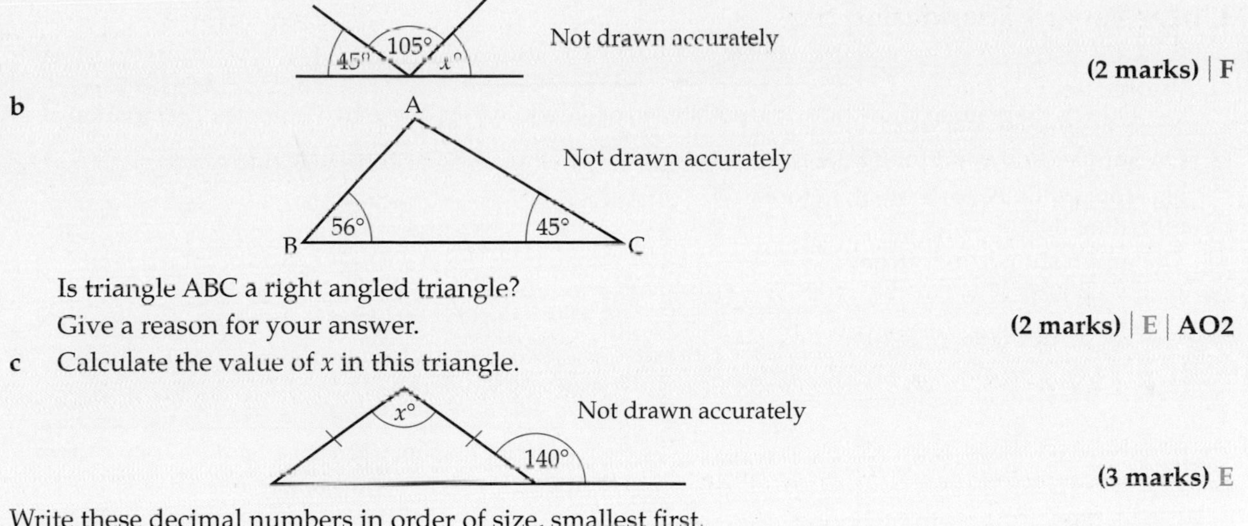

Not drawn accurately

(2 marks) | **F**

b Not drawn accurately

Is triangle ABC a right angled triangle?
Give a reason for your answer. **(2 marks)** | E | **AO2**

c Calculate the value of x in this triangle.

Not drawn accurately

(3 marks) E

8 Write these decimal numbers in order of size, smallest first.

0.52 0.09 0.5 **(1 mark)** | **F**

9 Kyle is visiting his brother in America. He arrives on 12th March.
He is allowed to stay in the country for 90 days only, including the day he arrives and the day he leaves.
What is the latest date that Kyle can leave America?
You **must** show your working. **(4 marks)** | **F** | **AO3** | **Funct.**

10 Here are two formulae.

$A = \frac{1}{2}bh$ $V = Al$

Work out the value of V when $b = 5$, $h = 4$ and $l = 6$. **(3 marks)** | **F** | **AO3**

11 Twenty students were asked what broadband provider they used.
Here are the results.

Orange	Tiscali	Tiscali	Orange	Sky
Sky	BT	Orange	BT	Orange
BT	Orange	Sky	Orange	Sky
Tiscali	BT	Sky	Sky	Sky

a Complete the table

Broadband provider	Tally	Frequency
Orange		
Sky		
BT		
Tiscali		

(2 marks) | **F**

b Draw a pie chart to represent this information.
You **must** show how you worked out the angles of your pie chart. **(4 marks)** | E

12 **a** Make an accurate drawing of this triangle.
You must leave in your construction lines.

8 cm 6 cm 12 cm

Not drawn accurately

(3 marks) | E

b Here are two identical right angled triangles.

Draw a diagram to show how it is possible to make a kite from these two triangles. **(2 marks)** | E | **AO2**

13 On Sunday Rob went for a bike ride.

He stopped twice for a break before returning home.

The graph shows his journey.

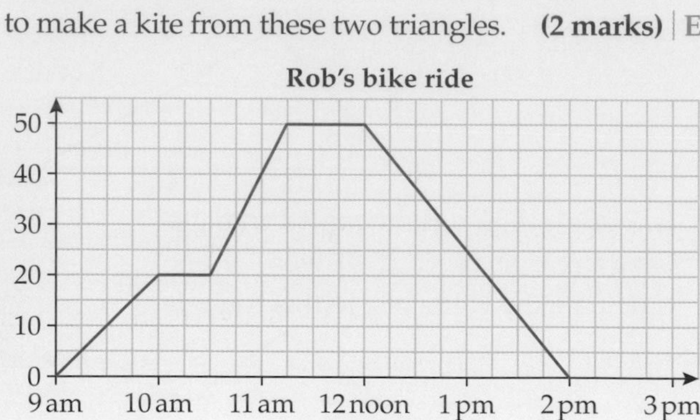

a At what time did Rob start his second break? **(1 mark) D** | **Funct.**

b Work out his average speed for the entire journey. **(2 marks) D** | **Funct.**

14 The frequency table shows the number of days absent due to illness of the teachers in a school in January last year.

Number of days absent due to illness	Frequency	
0	7	
1	4	
2	3	
3	4	
4	2	

a Write down the modal number of sick days taken in January last year. **(1 mark)** | D

b Calculate the mean number of days absent due to illness per teacher in January last year. **(4 marks)** | D

In June last year the mean number of days absent due to illness per teacher was 0.5 and the range was 7.

c Compare the number of days absent due to illness of the teachers in the school in January and June using the mean and the range. **(2 marks)** | D | **AO2** | **Funct.**

15 Anil has £200 in his savings account.

On Monday he pays an extra 10% into his savings account.

On Tuesday he takes out 10% of the money in his savings account.

Tick the box which describes the amount of money he has in his savings account on Wednesday.

Less than £200 ☐ Exactly £200 ☐ More than £200 ☐

Give a reason for your answer. **(2 marks)** | **D** | **AO2**

16 The table shows the adult : child ratios for different ages of children in daycare.

Age of children	Adult : child ratio
under 2 years	1 : 3
2 years	1 : 4
3–5 years	1 : 8

These are the ages of the children registered at a daycare centre.

Less than one year old	7
Over one year, but less than two	5
Two-year-olds	8
Three-year-olds	12
Four-year-olds	7

Work out the number of adults needed to care for these children.

You **must** show your working. **(4 marks** | C | **AO2** | **Funct.**

17 The table shows the waiting time, t minutes, that some patients waited in a doctors' surgery.

Waiting time, t (minutes)	Frequency
$0 \leq t < 10$	12
$10 \leq t < 20$	8
$20 \leq t < 30$	5
$30 \leq t < 40$	3
$40 \leq t < 50$	2

a Which class interval contains the median?
Explain how you worked out your answer. **(2 marks)** | C

b Explain why it is not possible to calculate the exact mean waiting time. **(1 mark)** | C | **AO2**

18 Abbie starts with this shape.

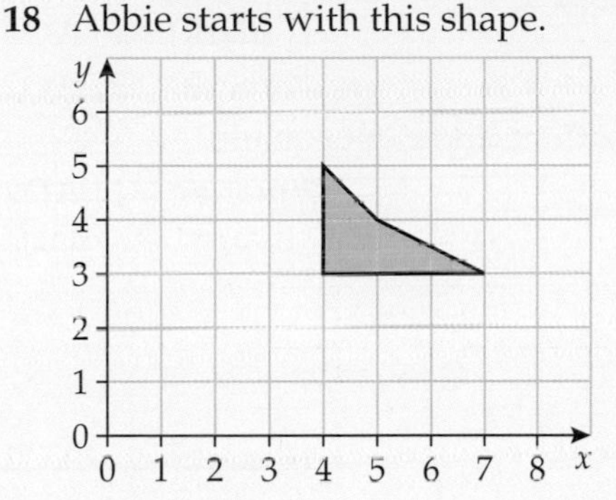

She transforms the shape to make this pattern.

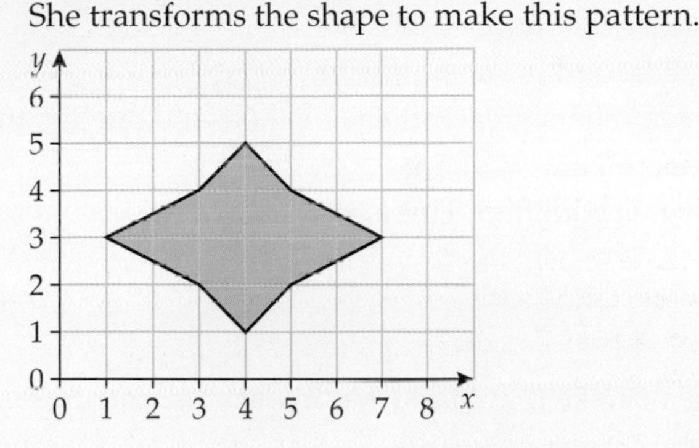

Describe the transformations she uses. **(4 marks)** | C | **AO3**

Linear Paper 2 Foundation tier — Calculator allowed

1 Sally goes shopping for clothes.

a She buy two pairs of jeans.

Jeans: normal price £24.99
Special offer!
Two pairs of jeans for £38

How much does she save using the offer? **(2 marks)** | G | **Funct.**

b Sally buys two jumpers.

Jumpers: normal price £14.50
Special offer!
Buy two and get the 2nd for half price

How much does she pay for the jumpers? **(2 marks)** | G | **Funct.**

c Sally buys four pairs of socks.

Socks: normal price £2.85
Special offer!
Buy three pairs and get the fourth pair free

How much does she pay for the socks? **(2 marks)** | G | **Funct.**

d Sally has £40 left.
She spends £12.50 on a pair of sandals.
She needs £3.80 for the car park.
She sees this advert for a jacket.

Jacket: normal price £32
Special offer!
25% off normal price

Does Sally have enough money to buy the jacket?
You **must** show your working. **(4 marks)** | D | **AO2** | **Funct.**

2 Here are some words that are used with angles and turning.

south **obtuse** **east** **acute** **west** **reflex** **right** **north**

Use one of these words to complete the following sentences.

a An angle that is less than 90° is called an .... angle. **(1 mark)** | G

b An angle that is exactly 90° is called a .... angle. **(1 mark)** | G

c An angle that is more than 180° but less than 360° is called a .... angle. **(1 mark)** | G

d If I face west and turn $\frac{1}{4}$ turn clockwise, I will be facing .... **(1 mark)** | G

e If I face east and turn $\frac{3}{4}$ turn anticlockwise, I will be facing .... **(1 mark)** | G

3 Here are four number cards.

3	8	5	6

a Use all four cards to make

i the largest even number possible **(1 mark)** | G | **AO2**

ii the smallest odd number possible **(1 mark)** | G | **AO2**

b How many numbers greater than five thousand but less than six thousand can be made?
You **must** show your working. **(2 marks)** | G | **AO2**

4 Marko is solving an equation. This is what he writes.

$x + 8 = 12$

$x = 12 + 8$

$x = 20$

a Explain the mistake that Marko has made. **(1 mark)** | G | **AO2**

b Work out the correct value of x. **(2 marks)** | G

5 The temperature in a city at midnight is −3°C.
By midday the temperature has risen by 9°C.
What is the temperature in the city at midday? **(2 marks)** | G | **Funct.**

6 Here is a list of numbers.

3 4 5 16 18 21 24 27

Write down two numbers from this list that are

a square numbers **(2 marks)** | G

b factors of 12 **(2 marks)** | G

c multiples of 9. **(2 marks)** | G

7 The pictogram shows the number of sandwiches sold in a canteen in one week.

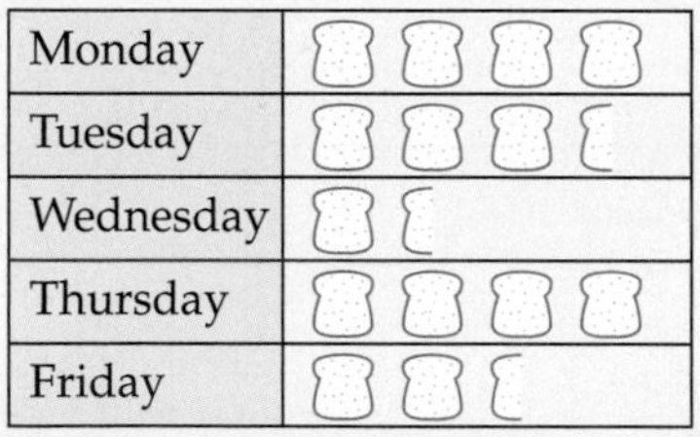

Monday	
Tuesday	
Wednesday	
Thursday	
Friday	

On Tuesday 35 sandwiches are sold.
What is the total number of sandwiches sold in this week? **(4 marks)** | G | **AO3**

8 a Make two copies of this diagram.

i On the first copy shade one more square to make a shape with one line of symmetry. **(1 mark)** | F

ii On the second copy shade two more squares to make a shape with two lines of symmetry. **(2 marks)** | F

b Amir says, 'The smallest number of triangles that need to be shaded in this shape to give it an order of rotational symmetry of 3 is 4.'

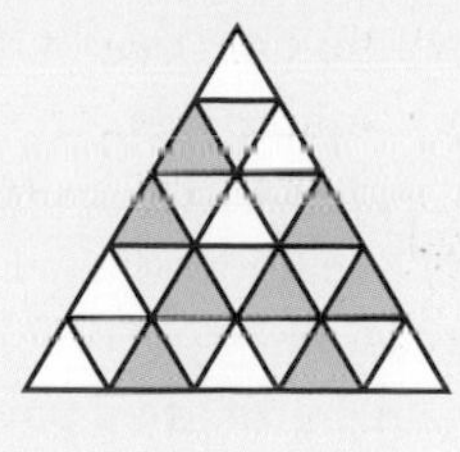

Is Amir correct? Explain your answer. **(2 marks)** | F | AO3

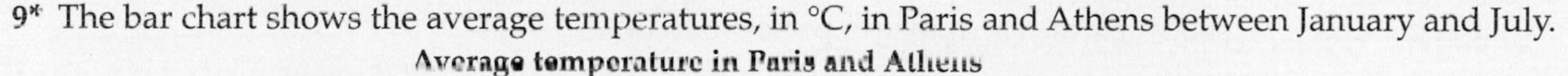

9* The bar chart shows the average temperatures, in °C, in Paris and Athens between January and July.

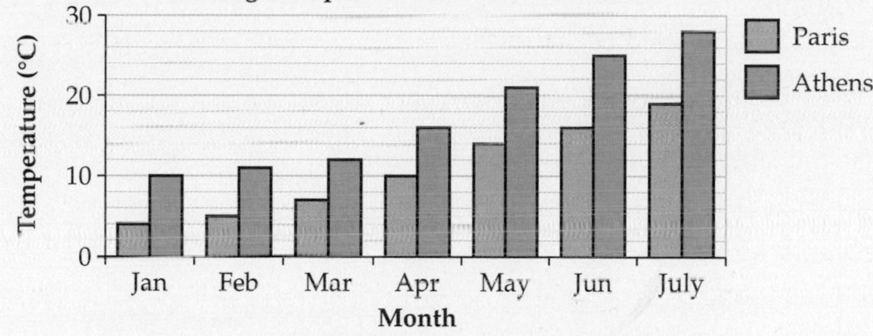

a In which month is the average temperature in Paris 16°C? **(1 mark)** | F | **Funct.**

b What is the average temperature in Athens in February? **(1 mark)** | F | **Funct.**

c What is the range in the average temperatures in Paris between January and July? **(2 marks)** | F | **Funct.**

d Shona wants to take her family on holiday in May. They like to sunbathe.
Use the chart above to decide if they should go to Paris or Athens.
Give reasons for your answer. **(1 mark)** | F | AO1 | **Funct.**

10 Kevin is working a night shift that lasts 12 hours.
At 11pm he has completed one third of his shift.

a How long has Kevin been working? **(2 marks)** | F | **Funct.**

b At what time will Kevin complete his shift? **(3 marks)** | F | AO2 | **Funct.**

11 This chart gives the driving distance in miles between different cities in the UK.

Hull				
251	**Cardiff**			
275	394	**Glasgow**		
41	246	208	**York**	
216	150	406	208	**London**

Calculate the distance in kilometres from Cardiff to York. **(3 marks)** | F | AO3 | **Funct.**

12 $m \Omega n$ means $4m + 3n$

a Work out the value of $2 \Omega 4$. **(2 marks)** | F

b If $5 \Omega y = 41$, work out the value of y. **(3 marks)** | E | AO2

c If $x \Omega 5 = 6 \Omega x$, work out the value of x. **(3 marks)** | D | AO2

13 Here is an isosceles triangle.

Show how two of these triangles can be joined to make

a a parallelogram **(1 mark)** | F

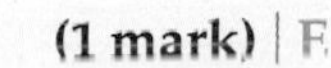

b a rhombus. **(1 mark)** | E

14 **a** Use your calculator to work out 5.4^3. **(1 mark)** | E

b Use your calculator to work out $\sqrt{6.25}$. **(1 mark)** | E

15 Write an expression for the perimeter of this rectangle in its simplest form.

$3x + 2$

$2x$

(3 marks) | E | AO2

16 The length of a footpath on a map is 18 cm.

The scale of the map is 1 : 25 000.

What is the length, in km, of the footpath in real life? **(3 marks)** | E | **AO1** | **Funct.**

17* The normal price of a leather sofa is £800.

If a customer pays cash they get one quarter off the normal price.

If a customer buys on credit, they pay a 20% deposit of the normal price, plus 12 monthly payments of £68.

What is the difference between the cash price and the credit price?

You must show your working. **(7 marks)** | E | **AO3** | **Funct.**

18 Greg has two four-sided dice, numbered one to four.

He rolls the dice at the same time.

His score is the difference between the two numbers.

a Copy and complete the table to show all the possible scores. **(2 marks)** | E

	1	2	3	4
1	0			
2				2
3				
4			1	

b What is the probability of getting a score greater than 1? **(1 mark)** | D

c At his school fete, Greg uses this game to raise money.

Each player pays 50p to play the game.

If a player gets a score of 3 they win £1.50

If a player gets a score of 2 they win £1

If 400 people play the game, how much money should Greg expect to make? **(5 marks)** | D | **AO3**

19 This is the floor area of a kitchen that is going to be tiled.

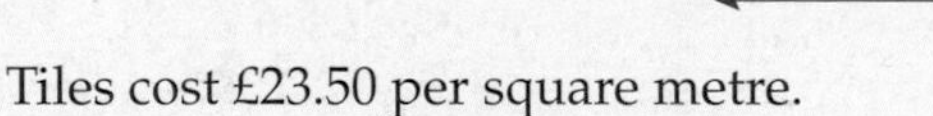

Tiles cost £23.50 per square metre.

They can only be bought in whole numbers of square metres.

Work out the cost of the tiles for the kitchen floor.

You **must** show your working. **(4 marks)** | D | **AO2** | **Funct.**

20 Expand and simplify $(x + 2)(x - 5)$ **(2 marks)** | C

21 a Copy and complete the table of values for $y = x^2 + 2x - 2$. **(2 marks)** | C

x	−3	−2	−1	0	1	2	3
y	1	−2		−2	1		13

b Draw a coordinate grid that goes from −3 to +3 on the x-axis and −4 to +14 on the y-axis.

On the grid, draw the graph of $y = x^2 + 2x - 2$ for values of x from −3 to + 3. **(2 marks)** | C

c Use your graph to write down the solutions to the equation $x^2 + 2x - 2 = 0$ **(2 marks)** | C

22 A is the point (2,1) and B is the point (6, 9).

a Work out the midpoint of the line AB. **(2 marks)** | C

b Calculate the length of the line AB.

Give your answer correct to one decimal place. **(2 marks)** | C | **AO2**

23* Show that $2(4p + 3) - 3(p - 2) = 5(p + 2) + 2$ **(4 marks)** | C | **AO3**